William Saunders, Robert Hodge, William Green

Observations on the Superior Efficacy of the Red Peruvian Bark

in the cure of agues and other fevers: interspersed with occasional

remarks on the treatment of other diseases by the same remedy

William Saunders, Robert Hodge, William Green

Observations on the Superior Efficacy of the Red Peruvian Bark
in the cure of agues and other fevers: interspersed with occasional remarks on the treatment of other diseases by the same remedy

ISBN/EAN: 9783337839659

Printed in Europe, USA, Canada, Australia, Japan

Cover: Foto ©berggeist007 / pixelio.de

More available books at **www.hansebooks.com**

OBSERVATIONS

ON THE

SUPERIOR EFFICACY

OF THE

RED PERUVIAN BARK,

IN THE

CURE of AGUES and other FEVERS.

INTERSPERSED WITH

OCCASIONAL REMARKS on the TREATMENT of
other DISEASES by the same REMEDY.

THIRD EDITION.

WITH CONSIDERABLE ADDITIONS.

And an Appendix, containing a more particular Account of its Natural History.

BY WILLIAM SAUNDERS,
M. D. F. R. S.

Member of the Royal College of Physicians in London, and Physician to Guy's Hospital.

Æque pauperibus prodeft, locupletibus ægre. Hor.

BOSTON
PRINTED BY ROBERT HODGE,
FOR WILLIAM GREEN.

INTRODUCTION.

THE superior efficacy of the Red Peruvian Bark has alone established its use, in preference to the common kind, and in opposition to the most interested views of dealers in the article of Pale Bark.

The intelligence I have received from every part of England, and from many parts, of the Continent of Europe, conveying the most unequivocal evidence of its active powers, would exceed many volumes ; I am therefore obliged to rest satisfied with assuring my readers that in no one instance where the Red Bark has been employed with judgment, has its superior efficacy been called in question. Its use, on its first introduction, was chiefly directed and confined to the cure of Intermittent Fevers, which, from their singular obstinacy, had resisted the common Bark ; farther experience has however clearly ascertained its great efficacy in other febrile disorders, and in cases of gangrene and scrophula, accompanied with a defective tone in the system. It may be admitted, that as the Red Bark contains more resinous parts, and is more active, that greater caution is necessary in the use of it, and that an injudicious application of it may prove hazardous to the constitution ; this argument will however apply to all remedies in their most active and perfect forms.

I am now employed in collecting facts from the best authorities, and comparing them with my own observations in order to ascertain with precision the real operation of Bark on the habit, and the particular circumstances and periods of diseases which either justify or condemn the use of this valuable remedy.

Since

Since the publication of the second Edition of this Treatise, I have not been disappointed in any one instance of curing Agues, and even some of the most complicated and unfavourable nature, by the use of a cold infusion of the Red Bark in Water, a preparation of it which always sits easy and light upon the stomach, and which is sufficiently impregnated with the powers of this medicine, to answer every purpose to be obtained by Peruvian Bark.

I believe a cold infusion of the Pale Bark has never been considered as sufficient to cure Intermittent Fevers.

I have procured, by the favour of my learned friend Dr. Simmons, some important information regarding the natural history of the Red Bark, which I have inserted in this Edition of my Treatise; we are now sufficiently encouraged to hope for a liberal and regular supply of this kind of Bark from its growing in the province of Santa-Fe, through which a large river flows which empties itself into the harbour of Carthagena: for other particulars I refer you to the Doctor's Letter, which is replete with useful information.

My friend, Mr. Aikin, of Warrington, whose works in matters of taste, as well as science, have been so universally admired, and who is now preparing for the press a new edition of that excellent book on the Materia Medica, by Dr. Lewis, confirms the truth of my experiments on the Red Bark. I have thought proper, in this edition, to give an extract of this Letter, from an author in whose accuracy and candour so much confidence may be reposed. I lately saw a letter from a Physician in St. Domingo to his friend in London, informing him that a Peruvian Bark of a red colour, and in general of a larger size, was lately introduced into that island, and which had proved more efficacious than the pale Bark; he advised his friend to speculate largely in the article, and assured him that he might be supplied with it next year in any quantity;

quantity; he fuppofed it was from a new foreft of trees which had been lately difcovered.

My friend, Dr. Lifter, now at Paris, has been able to procure me from the collection of M. Jofeph de Juffieu, the different Specimens of Bark mentioned in Dr. Simmon's Letter, and particularly one Specimen, which, by the defire of the Spanifh Minifter for the American department, was prefented to the Royal Medical Society at Paris, by Don Cafimir Ortoga, Profeffor of Botany at Madrid. I was lately requefted to procure a quantity of Red Bark of the beft quality for Le Compte de Carburi, Phyfician to the Count D'Artois, fo that the ufe of it will probably foon become general on the continent.

There is fome danger, from the avarice of dealers, of its being adulterated, more efpecially, however, in the form of powder, a circumftance which may bring it into difcredit, I have therefore taken fome pains in collecting fpecimens of it, and arranging them in the order of their goodnefs, chiefly for the benefit of the young Gentlemen who attend my Lectures. The many applications to me on this fubject, rendered this plan neceffary, the Red Bark being hitherto little known by Apothecaries in the country. The efficacy of this remedy is fo well eftablifhed, that it is a matter of very little confequence, if Botanifts fhould ftill fuppofe that there are not Data fufficient to determine whether it be from the fame fpecies or not, with the Cinchona Officinalis of Linnæus. My friend, Dr. Hope, Profeffor of Botany, in the Univerfity of Edinburgh, in a Letter to me, fays, " If this Red Bark,
" which feems to be more efficacious than the Com
" mon Bark, can be had ; that is the great point, and
" whether it be, or be not, the Cinchona Officinalis
" of Linnæus, is but a fecondary, and much lefs im
" portant confideration ; however, I think your opi
" nion the moft probable ; your Treatife will, I think,
" have

" have the effect of making the public attend to this
" matter, and of rendering the ufe of it more general."

Since the following pages went to the prefs, I have
feen fome exceeding good Red Bark, imported by a
Spanifh Merchant, a confiderable part of which was
as fmall as the Quilled Bark in common ufe, but it
ftill preferved its Rednefs in that form, approaching,
however, more to the colour of Cinnamon ; it was
evidently more compact and heavy than common
Quill Bark, and appeared extremely refinous, its ex-
terior coat, thin, whitifh, and rough ; it broke brittle, and gave evident proofs of its being the Quin of
the larger Red Bark, which was in the fame cheft : this
conveys the agreeable information of its being either a
variety of the Cinchona Officinalis of Linnæus, or perhaps a diftinct fpecies, fo that the trees may not be
confidered as in danger of being deftroyed by the introduction of this Bark. A difference of colour in
the fame fpecies is often influenced by age, foil, or expofure.

O F

R E D B A R K.

I HAD long fufpected, that the Peruvian Bark in common ufe, was very inferior in power and efficacy to that recommended by the early writers on the fubject; but more efpecially by our countrymen, MORTON and SYDENHAM, in whofe works the medical virtue of this drug, in Intermittent and other Fevers are extolled as little fhort of infallibility. In their time the Quill Bark, (at leaft fuch as is now in ufe), was not mentioned; their cotemporary writers on the *Materia Medica*, evidently defcribe the Peruvian Bark of that period, as of a larger kind, of more compact pieces, and of the colour of the ruft of iron, which marks are very expreffive of the Red Bark; the innermoft coat of which has an ochrey appearance, and its refinous or middle layer refembles very much the *Lapis Hæmatitis* *. M. LA CONDAMINE expreffed his furprife, when he was told by Mr. THOMAS BLACHYNDEN, Director of the Englifh South Sea Company, at Panama, that the writers on Pharmacy and *Materia Medica* in England had preferred the Small and Quill Bark, while the inhabitants of New Spain,

held

---

* Cortex craffitie mediocri, foras fcaber, paululum canefcens, aliquando mufco obductus, intus lævis, coloris ferruginei, faporis acris et intenfe amari. Autumno colligitur et cortex circumcirca delibratur, tam trunci quam ramorum.

Dalei Pharmacologia, p. 201. *Anno* 1692.

held the larger Bark in higher eſtimation. * I can only explain ſuch a preference from this circumſtance,
that the larger pieces of Bark, which were imported
into this country along with the ſmaller Bark, were
either of a fibrous, ſpongy, or ligneous texture ; or
probably, damaged by moiſture, and taken from the
decayed trees.

Druggiſts have ſometimes ſuggeſted this as an argument againſt the uſe of the Red Bark, and many
probably endeavour to oppoſe its introduction, until
the quantity in the markets be greater, and more
equally divided among them, or until they have been
able to diſpoſe of the large quantity of common Bark
they have in their ware-houſes.

The taſte and flavour of the Red Bark is more difficultly evolved, and is therefore at firſt not ſo obvious
from the cloſeneſs of its texture, and from the reſinous
coat being ſo well defended and incloſed between two
other layers. It is evidently heavier than any other
kind of Bark, and ſeems to have been prepared and
dried with greater attention, its original appearance
and form being better preſerved.

I think it probable from a more attentive view of
the ſubject, that it may be the Bark of the trunk, or
larger

---

* They commonly reckon three ſpecies of the Quinquina, though
ſome make four ; the white, the yellow, and the red : I was told at
Loxa, that theſe three kinds differ in their virtue only ; the white
having ſcarce any virtue, and the red excelling the yellow.
The trees from whence the firſt Bark was taken, which were very
large, are all dead, having been entirely ſtripped, which infallibly
kills them when they come to be old. Experience has ſhewn, that
ſtripping kills ſome of the young ones alſo, but the greateſt part eſcape.
For this operation they uſe a common knife, which they hold in both
hands ; the barker ſticks it into the bark as high as he can reach, and
ſo draws it downwards as low as he can. It does not appear that the
trees which grow where the old ones ſtood, have leſs virtue than they,
the ſituation and ſoil being the ſame ; the difference, if there be any,
may ariſe, perhaps, from the different ages of the trees. Few but
young ones are now to be met with : I do not remember to have
ſeen any much thicker than my arm, or above 12 or 15 feet high ;
thoſe which are cut young, ſhoot forth new branches from the ſtumps.
I was

larger branches of the tree, and I am the more con-firmed in this opinion by the ideas of my friends Dr. Withering and Dr. Fothergill, conveyed in their letters to me, in which they obferve, that the effential and active parts of the Oak Bark are more entire, and in larger quantity in the trunk and larger branches, than in the twigs or fmaller branches, which are com-paratively of an imperfect growth ; perhaps the Small and Quilled Bark may be procured from younger trees, not yet arrived at their full maturity, and therefore yielding a Bark of a weaker quality *.

I am juftified very much by the analogy of other Barks, and by the influence of foil and expofure, in changing the appearances, and even of affecting the virtues of the fame fpecies of vegetables, to conclude, that the diverfity both in fize and colour of the Red Bark from the common Peruvian Bark, may depend

B either

I was informed at Loxa, that heretofore they preferred the coarf-eft Bark, and laid it by as a rarity, but now the fineft is moft efteem-ed : The merchants may poffibly find their account in it, as it takes up lefs room in packing. But a Director of the Englifh South Sea Company at Panama, through which all the Quinquina that comes to Europe muft pafs, affured me, that the preference given at prefent to the fine Bark, is in confequence of feveral chymical analyfes and experiments which have been made on both forts in England. It feems probable, that the difficulty of thoroughly drying the large coarfe Bark, and the humidity it is naturally apt to contract and re-tain, has helped to bring it into difrepute. Vulgar prejudice will have it, that to lofe nothing of its virtue, the tree fhould be barked in the moon's decreafe, and on the eaft fide. Thefe circumftances, as alfo its being gathered on the mountain of Cajanuma, were certified by a Notary in 1735, where the Marquis de Caftlefuerte procured a quan-tity of Quinquina from Loxa, to carry to Spain on his return.
Memoir de l' Acad. d. S. 1738.

* Mr. Arrot, a Scotch Surgeon, who had gathered the Bark in the place where it grows, fays, that the fmall curled Bark fo much efteem-ed in England, is the Bark of younger trees, which frequently reco-ver the Barking, while the older trees never do. This affords a ftrong proof, that the early bark introduced into Europe was of the larger kind, and from the older trees, while the difficulty of procur-ing it, has been the means of introducing a fmall and younger Bark. Mr. Arrot, and every other Traveller agrees in preferring the Red Bark to any other.
Encyclopædia. Cortex. Vol. I.

either on the largeness of the branches, or on other circumstances not necessarily implying a specific difference in the tree.—On comparing the larger Bark of the Oak with its twig Bark even collected from the same tree, I find the former of a reddish colour, while the latter is much paler, the roughness of the coat of the larger Oak Bark and its general appearance may be compared to that of the Quill Oak Bark, as the Red Bark is to the common Peruvian Bark ; and I have not only learned from consulting Tanners on this subject, that the larger Bark is superior in its powers as an astringent; but I have found by comparing infusions of both, and submitting them to the most decisive experiments regarding their astringency, by adding to them solutions of iron, that the precipitates were of a blacker colour and in greater quantity, from the larger and more compact pieces of Bark, than from the smaller twig Bark.—I have convinced many medical gentlemen, who have seen my specimens of the Oak Bark, that they tend very much to illustrate and confirm the opinion that the Red Bark is the *Cinchona Officinalis* of *Linnæus.*

The following description of the Peruvian Bark taken from Pomet is likewise a further confirmation of this doctrine ;—" The *Kinquina* is the bark of a " tree that grows in *Peru*, in the province of *Quitto*, " upon the mountains near the city of *Loxa.* This " tree is almost the size of a Cherry-tree, the leaves " round and indented ; it bears a long reddish flow- " er, from whence arises a kind of pod, in which is " found a kernel like an almond, flat and white, " cloathed with a flight rind ; that Bark which " comes from the trees at the bottom of the moun- " tains is thicker, because it receives in more nourish- " ment from the earth ; it is smooth, of a whitish yel- " low without, and of a pale brown within. That " which comes from trees on the top of the moun- " tains is abundantly more delicate ; it is uneven,
" browner

" browner without, and of an higher colour within ;
" but the trees which grow on the middle of the
" mountains, have a Bark yet browner than the other
" and more rugged. All thefe Barks are bitter, but
" that from the trees at the bottom of the mountains,
" lefs than the others.

" It follows from hence, that the Bark of the leaft
" virtue, is that which grows in the loweft places, be-
" caufe it abounds more with earthy and watry parts,
" than that which grows high, and which for the
" contrary reafon is better ; but the beft of all is that
" which grows in the middle of the mountains, be-
" caufe it has not too much or too little nourifhment.
" There is another kind of Bark which comes from
" the mountains of *Potofi*, and is browner, more aro-
" matic, and bitterer than the former, but abundant-
" ly fcarcer than any of the reft.

" The conditions or qualities we ought to obferve
" in the Bark, are, that it be heavy, of a firm fub-
" ftance, found and dry. Beware of fuch as is rot-
" ten, and will fink in water prefently, and that flies
" into duft in breaking, or is dirty and unclean, as it
" happens to be fometimes ; but make choice of fuch
" as is in little thin pieces, dark and blackifh without,
" with a little white mofs, or fome fmall fern-leaves
" fticking to it, reddifh within, of a bitter and difa-
" greeable tafte ; and refufe that which is full of
" light chives when broke and of a ruffet colour, and
" take care that there be not feveral pieces of wood
" mixed with it, which you have more of fometimes
" than the Bark.

" This was brought firft into France in the year
" 1650, by the Cardinal Lago, a Jefuit, who having
" brought this from Peru, it was had in fuch vogue
" in France, as to be fold weight for weight at the
" price of gold."

I have fince the publication of the firft edition of
this treatife, extended my enquiries into other countries,

tries, from which I am convinced that there cannot a doubt be entertained of its being the *Cinchona Officinalis*. I have a specimen of the Red Bark which was given me by Mr. Babbington, the Apothecary of Guy's Hospital—it contains in it a branch of Quill Bark, exactly as it was imported.

I have seen some specimens of Red Bark so very large, that they contained a great proportion of woody part, and therefore less fit for use than those of a more moderate size ; indeed from having made such frequent experiments on this subject, I am able with great precision to ascertain the comparative quantity of resin in any two pieces of Bark from their external appearance.

It has been suggested by some, that the Red Bark resembled much the mahogany Bark ; but having examined that Bark, and having conversed with persons to whom it is extremely familiar, I am persuaded that there is no foundation for the opinion of its being the same. If future naturalists, by having better opportunities of investigating this part of our subject, should be persuaded that it is the Bark of a tree of a different *genus*, or *species*, from the *Cinchona Officinalis*, such a discovery cannot invalidate the proofs of its superior efficacy ; and I should have been happy had it been found to be the production of any of our colonies, instead of its being as yet known in Europe only as a native of South America. Several very intelligent men, who were disposed to think it the Bark of a different tree, immediately changed their opinion, from examining it in tincture, decoction, or infusion, in which forms it conveys the genuine taste and flavour of the common Peruvian Bark, under the appearance of a much stronger impregnation. I suspect that we have been long in error by judging chiefly of the goodness of Peruvian Bark, from the colour of its external coat. I have seen some specimens of Red Bark extremely rich in resinous parts with a very

white

white coat, but whose inner layers were compact, and of a dark red or ochry colour. I have examined twenty chests of this Red Bark in the very state in which it was imported, and there is always found a very considerable proportion of Quill Bark amongst it.

If the execution of this work was equal to its importance, it would challenge the attention of the public, in a degree far above most medical subjects. This will be unquestionably admitted by those who have been eye witnesses to the malignancy and fatality of intermittent and remittent Fevers, in every part of the globe, but more especially in warm climates ; this fatality is by no means owing to the ignorance or unskilfulness of the practitioners in those countries, but to the inefficacy of the common Bark in general use.

The numbers who fall a sacrifice to the epidemic and seasoning Fevers of warm climates, are admitted infinitely to exceed those who are destroyed by the enemy. In almost all the dangerous Fevers which occur in our East and West India settlements, the Bark is a principal remedy. I think it therefore an object of the greatest national importance, that our fleets and armies should be liberally supplied with this Bark, which will seldom or ever disappoint them.

I have been told by a druggist, that its great activity rendered it a dangerous remedy ; my answer was, that wine mixed with water was much safer in the hands of an unskilful practitioner, than wine alone, but that did not prove that wine was not a better cordial than water. The same reasoning may be applied to prove, that weak and decayed remedies, by being much milder in their operations, are therefore preferable to such as are more perfect of their kind. I believe the general, and best founded complaint is against the want of power and efficacy of Bark, and not that it is too powerful and active. In proof of this, I refer my readers to the letters annexed, which demonstrate, that

such

such was the stubbornefs and obstinacy of the intermittent Fevers of the present year, even in this country, that they resisted common Bark, and only gave way to the Red Bark.

Notwithstanding I formed very early a favourable opinion of this Bark, yet it fell far short of that which I am justified in maintaining, from the collected evidence of so many gentlemen in distant and remote parts of the country. Many letters which I have received, are written with such zeal in favour of its superior efficacy, that they could only be dictated by the strongest conviction, arising from extensive and diligent observations.

I have persuaded many of my medical friends to use the Red Bark in our foreign settlements, and I shall take pleasure in communicating to the public the result of their observations as soon as I am favoured with them. A more powerful Bark is particularly desirable in those countries, where the violence and danger of the paroxysm is so great. In the following observations, I have confined myself very much to the use of the Red Bark in febrile diseases, but I am in possession of many facts in proof of its superior powers in other diseases, in which the common Peruvian Bark has been found useful.

The introduction of the Red Bark into this country was the effect of chance. In the year 1779, a Spanish ship from Lima, bound to Cadiz, was taken by the Hussar frigate, and carried into Lisbon; her cargo consisted chiefly of this Bark, some part of which was immediately imported into this country, and a considerable quantity was bought at a very low price at Ostend, by some of our London Druggists. The boxes in which it was brought to Europe were of the same kind as those in which the common Peruvian Bark was contained, and all sold by the general title of Quinquina. The Druggists in whose hands the Red Bark at first was, found it difficult to dispose of it, it, its apearance was so very unlike that of common bark; at last they offered it by way of trial to such Apothecaries as reside in counties where agues are frequent; the succefs attending its ufe foon convinced them of its fuperior efficacy. It was early introduced into the hofpitals, and its greater powers became univerfally acknowledged. It has continued ever fince in general ufe in the Hofpitals of St. Bartholomew, St. Thomas, Guy, and the London. The reputation, therefore, of the Red Bark ftands better eftablifhed, and is fupported by the concurring teftimony of more Phyficians, than that of any other article of the *Materia Medica.* I am affured by every Druggift with whom I have converfed on the fubject, that the demand for it in this country is prefling and general. I am likewife informed, that the markets may be well fupplied with it; and, as it is no longer in the hands of a few dealers, the prejudices of the Druggifts have fubfided, and I have lately heard nothing of its deftructive qualities, which were faid to have arifen from its fuperior powers.

Being highly fenfible of the difficulty of eftablifhing fuch facts, either on the effects of remedies, or on any branch of medicine which regards the animal œconomy, I have folicited the opinion of many ingenious and attentive practitioners, who from their fituation have had frequent opportunities of trying the Red Bark. This caution appeared the more neceffary, becaufe I am well perfuaded that the love of novelty, and too great a credulity in admitting facts on very doubtful authorities, have corrupted medicine more than any other fcience, and proved more injurious than the moft abfurd and fanciful theories, the errors of which are eafily detected.

O F

T H E Red Bark is in much larger and thicker pieces than the common Peruvian Bark. It evidently confifts of three diftinct layers. The external, thin, rugged, and frequently covered with a moffy fubftance, and of a reddifh brown colour *. The middle, thicker, more compact, and of a darker colour. In this appears chiefly to refide its refinous part, being extremely brittle, and evidently containing a larger quantity of inflammable matter than any other kind of Bark.

The innermoft has a more woody and fibrous ap-pearance, and is of a brighter red than the former.

The intire piece breaks in that brittle manner defcribed by writers on the *Materia Medica*, as a proof of the fuperior excellence of the Bark.

In reducing it to powder, the middle layer, which feems to contain the greateft proportion of refin, will not give way to the peftle fo eafily as the other layers; this fhould be particularly attended to when it is ufed in fine powder. Its flavour is chiefly difcoverable either in powder or folution, is evidently more aromatic, and has a greater degree of bitternefs than the common Bark.

OF ITS CHYMICAL AND PHARMACEUTICAL HISTORY.

EXPERIMENT I.

TO an ounce of Red Bark, reduced to a fine powder, were added fixteen ounces of diftilled water; and

* I have lately feen fome very good Red Bark whofe external coat had a white appearance, though its internal furface is of a deep red colour, extremely refinous, compact, and heavy.

and after remaining together twenty-four hours in a Florence flask, the liquid was carefully filtered. The same experiment was made with the Peruvian Bark commonly in use.

The colour of the two infusions was very different, that made with the Red Bark being much deeper. The taste and flavour of the infusion of the Red Bark were considerably more powerful than of the other. In the opinion of many gentlemen who tasted the infusions, the cold infusion of the Red Bark was more sensibly impregnated than even the strongest decoction of the common Bark.

EXPERIMENT II.

TO two ounces of the cold infusion of Red Bark, were added twenty drops of the *Tinctura Florum Martialium*. It immediately became of a darker colour, soon lost its transparency, and after a short time precipitated a black powder.

EXPERIMENT III.

TO two ounces of the cold infusion of the common Bark were added twenty drops of the *Tinctura Florum Martialium* in the same manner as to the other. It retained its transparency some time, and afterwards became of a dark colour, but there was no precipitation from it as from the last.

EXPERIMENT IV.

TO an ounce of Red Bark, reduced to a coarse powder, were added sixteen ounces of distilled water, and after boiling until one half was evaporated, the liquid while hot was strained through a piece of linen. The same experiment, under similar circumstances, was made with the common Bark. The superior taste

C   and

and flavour of the decoction of the Red Bark was equally obfervable with that of the infufion. The decoction of the Red Bark, in cooling, precipitated a larger quantity of refinous matter than the decoction of the common Bark. The difference of colour was likewife very diftinguifhable.

EXPERIMENT V.

To an ounce of Red Bark, reduced to a coarfe powder, were added eight ounces of proof fpirit, and, after ftanding a week together, the Tincture was filtered.

The fame experiment, under fimilar circumftances, was made with the common Bark. The Tincture of the Red Bark, both when tafted by itfelf and under precipitation by water, had more flavour and tafte than that of the common Bark.

The Tincture from the Red Bark is of a much deeper colour than the other.

EXPERIMENT VI.

To each *refiduum* of the above Tinctures were added eight ounces of proof fpirit, which were infufed in a moderate fand heat for the fpace of twenty-four hours, and afterwards allowed to remain together a week, occafionally agitating them. The Tinctures were then poured off, that of the Red Bark evidently appearing to be the ftrongeft.

The Tinctures both of Experiments V. and VI. were by a gentle heat evaporated to the confiftence of a refinous extract.

The extract from the Tincture of the Red Bark was of a fmooth, homogeneous appearance, not unlike the Balfam of Peru, when thickened : The flavour and tafte of the original Tincture were intirely preferved in it.

The extract from the Common Bark had a very different appearance. It seemed coarse and gritty, and by no means so characteristic of its original Tincture.

The quantity of extract procured from the Red Bark was considerably greater than from the same quantity of common Bark ; but as the *residuum* of neither was rendered entirely inert, the absolute quantity could not be ascertained. *

EXPERIMENT VII.

A tea spoonful of each of the Tinctures, prepared by Experiment V. was added to two ounces of water; the resinous precipitation from the Red Bark was not only more copious, but fell more quickly to the bottom of the glass than that from the other, and yet what remained still dissolved in the water, was infinitely more in the Red Bark than in the common Bark, so far as we could judge from the taste and flavour of both.

EXPERIMENT

---

* To 26 lb. of Red Bark were added 26 gallons of proof spirit, after remaining together for some time the Tincture was poured off, and submitted to a distillation in a water bath, the quantity of spiritous extract obtained was 12 lb. and a half ; a quantity of water being poured on the *residuum* of the Tincture, the watery extract obtained was 4 lb.

In another experiment with 30 lb. of Red Bark, of an inferior quality, treated in the same manner as the former, only 11 lb. and a half of spiritous extract was procured, and 4 lb. and a half of watery extract.

To the same quantity of the best Peruvian Bark hitherto in use gives from 6 lb. and a half to 7 lb. and a half of spiritous extract.

It may be proper to observe, that the facts here mentioned are on the authority of a very eminent druggist, who had accurately marked the quantity of extract obtained by the usual process from a given quantity of Red Bark at two different trials; the reader will likewise observe, that although the spirit employed for making the Tinctures may have been saturated with resin, yet a fresh quantity was not poured on the residuum, which by extracting the whole resin would have yielded a larger proportion of resinous extract, and consequently left little or nothing for the watery extract. Though these experiments were not made with any view to a philosophical purpose, yet I am sufficiently convinced of their accuracy ; they are more conclusive than experiments conducted on a much smaller scale.

EXPERIMENT VIII.

In imitation of the experiments of my ingenious friend Dr. Percival, I added to two ounces of the watery infusion of each Bark a few drops of the *Sp. Vitriol, ten.* The acid loft its tafte more in the infufion of the Red, than in the common Bark, fo that there were more obvious appearances of its being neutralized.

EXPERIMENT IX.

A decoction of both Red and Common Peruvian Bark was prepared by taking an ounce of each and boiling them in a pint and a half of water, to one pint ; the former had greatly the fuperiority in ftrength and power, as mentioned in a preceding Experiment. A pint of frefh water was added to each decoction ; the boiling ftill continued till that quantity was evaporated. The decoction of the common Peruvian Bark feemed gradually to lofe its fenfible qualities, while that of the Red Bark ftill retained its own.

The fame quantity of water was added as before to each, and the decoction repeated until a gallon of water was exhaufted ; at the expiration of which time, the common Peruvian Bark was rendered almoft taftelefs ; the Red Bark ftill retained nearly its former fenfible qualities. This experiment proves that the common practice of boiling the Bark is hurtful to its powers.

By my defire Mr. Skeete, a very ingenious and attentive young gentleman from Barbadoes, and a ftudent of medicine in Guy's Hofpital, made feveral Experiments in order to afcertain the comparative antifeptic power of Red Bark, with the common Peruvian Bark ; and he found that the infufion of Red Bark preferved animal matter much better, and for a longer

ger time, than the infusion, or even decoction of the common Bark, indeed, the decoction of common Bark, after its powdery part had subsided, was less bitter, and preserved animal matter for a shorter time than the infusion of the same Bark. His experiments were conducted with great accuracy, and the result of them were submitted to the examination of many gentlemen at Guy's Hospital.

The conclusions to which the above experiments evidently lead, are,

First, That the Red Bark is more soluble than the Peruvian Bark, both in water and spirit.

Secondly, That it contains a much larger proportion of active and resinous parts.

Thirdly, That its active parts, even when greatly diluted, retain their sensible qualities in a higher degree than the most saturated solutions of common Bark.

Fourthly, That it does not undergo the same decomposition of its parts by boiling, as the common Peruvian Bark.

Fifthly, That the Red Bark is more astringent than the common Peruvian Bark.

Sixthly, That its antiseptic powers are greater; as an additional proof of this it may be proper to observe here, that both its cold infusion and decoction preserved entire their bitter and other medicated powers in the month of June, in the Elaboratory of Guy's Hospital for five weeks, and perhaps for a much longer time, while a decoction of common Bark gave evident marks of a change in a few days. In the decoction of Red Bark, the powder, which is separated during the cooling of it, remains intimately diffused through the liquor, which therefore continues loaded and turbid when at rest. In the decoction of common Bark, the powder quickly subsides to the bottom, the Red Bark therefore contains in it a large proportion of mucilaginous parts, such as have been proposed

poſed by the late Dr. Fothergill, to be added to the decoction of the common Peruvian Bark, in order that it may remain turbid when at reſt, and thereby that its reſinous parts be more perfectly ſuſpended in the body of the liquor. It is obvious, that this circumſtance will favour exceedingly the action of the ſtomach upon it.

The advantages therefore to be expected from the Red Bark cannot be obtained from any quantity of common Bark, the beſt common Bark, compared with the Red Bark appears inerte and effete.

All the above experiments were executed in the preſence of ſeveral Gentlemen.

- I was led more particularly to proſecute this ſubject, from an opinion that the Red Bark might ſo impregnate cold water by infuſion, as to cure Intermittent Fevers with more certainty than could be done even by the decoction or powder of common Bark: The ſenſible qualities which appear from the above Experiments, being ſo much greater, in the cold infuſion of the one than in the decoction of the other.

It cannot I think be denied, that the Experiments above related, and which have been executed and frequently repeated with great accuracy, ſufficiently prove, that the Red Peruvian Bark, exceeds the other in its ſenſible qualities, and that it contains a much larger proportion of thoſe reſinous and active parts on which the power and efficacy of Bark have been by all writers on the practice of medicine and *Materia Medica* believed to depend.

OF THE GENERAL OPERATION OF BARK ON THE HUMAN BODY.

T H E following remarks are intended to apply to the Peruvian Bark, generally in uſe; but I am certain that

that the effects enumerated are found to be produced in a much higher degree by the Red Bark.

The cold infusion of Bark seems evidently to promote both appetite and digestion, it increases the tone and action of the stomach, by which the gastric liquor, the great *Menstruum* of our solid aliment, is more perfectly prepared.

In most cases, the Bark rather promotes costiveness, the common effect of strong and vigorous intestines. In very large doses, however, it generally proves purgative, but this effect ceases after a short time.

It renders the pulse stronger and fuller in health, and in most diseases unaccompanied with Fever.

In Low and Malignant Fevers, and more especially under remission, it renders the pulse stronger and even slower.

In particular circumstances of Fever marked with debility and a tendency to remission, though of a very irregular type, it diminishes febrile heat.

It encreases the animal heat and aggravates every symptom, in Fevers accompanied either with much local inflammation, or a general inflammatory diathesis, which is strongly indicated by the pulse, the manner of the original attack, the want of due freedom in the secretions, and the painful and oppressive exercise of every function.

It checks profuse and colliquative discharges, especially those by the skin, while it does not seem to diminish insensible perspiration, or other natural evacuations.

It checks every tendency to putrefaction or gangrene, occurring under circumstances of debility, but it may promote both, if injudiciously employed, while the action of the system is too violent, or the inflammatory diathesis is too prevalent. It seems more reasonable to refer its action, as an antiseptic, to its tonic
power

power on the moving fyftem, than to any primary action on the animal fluids.

Perhaps this doctrine will apply in explaining the hiftory of remedies ufed in the Scurvy, a difeafe invited and favoured by every means which can induce debility, and evidently preceded by fymptoms of a diminifhed *vis vitæ* which neceffarily lead to others that in a fecondary manner only take place in the animal fluids *.

It promotes under many circumftances a favourable fuppuration, and improves the nature of fanious and ichorous difcharges.

Its action here can only be explained from its general tonic power, for either general or local debility retards fuppuration, and favours the obftinacy of ill-conditioned ulcers.

Upon the fame principle its power of promoting the generation of true *pus* in the fmall-pox may be explained.

The period of debility is that only in which the Bark fhould be employed.

I have feen patients under the moft confluent Small Pox require Bark in the progrefs of fuppuration, and yet, in the more advanced ftate of the difeafe, the fame perfons have been faved by the feafonable and repeated ufe of the Lancet in the fecondary Fever, which attacked with frefh rigors and inflammatory fymptoms of a true Peripneumony.

In delicate and irritable habits, which feem more efpecially to favour fcrophulous affections, and which produce inflammation of a peculiar nature, that gives way fooner to tonics than evacuants, the Bark has been found the beft remedy: this may probably admit of the explanation already given on the fubject of Scurvy.

OF

* An Enquiry into the fource from whence the Symptoms of Scurvy and of putrid Difeafes arife, by Dr. Milman.

OF ITS USE IN THE CURE OF INTERMIT-
TENT FEVERS.

WHAT I have to offer on this subject is the result of careful, and diligent obfervation, totally un-
connected with prejudice of any kind, in favour of
any particular theory, or a blind attachment to fyftem.
Notwithftanding the cautious and timid practitioner
has very generally forbid the ufe of Bark until evacuations fhould have been made ; I am very well perfuaded from obfervation, that in Intermittent Fevers,
fuch as rage and are endemic, particularly in low and
marfhy fituations, and fuch as frequently occur on the
banks of the Thames, and the lower parts of this me-
tropolis, the Bark cannot be given too early ; the ufe
of emetics or purgatives, as preparatory, is not only
unneceffary, but in fome cafes productive of greater
debility, and therefore to be avoided.

The doctrine of concoction, however juft it may be
in continued Fevers, and in fome cafes of inflammation, does not apply in Intermittents produced by the
Miafmata of low and fwampy grounds, and which afford the principal fource of agues in the environs of
this city.

In this opinion I am confirmed by the teftimony of
Dr. Cleghorn and others; who frequently found it
neceffary to give it on the firft acceffion of the difeafe,
in order effectually to obviate, or weaken the return
of a fecond paroxyfm, which in many cafes would in-
evitably have proved fatal ; and Dr. Lind has very
properly obferved that fuch fymptoms which have
been attributed to Bark, are rather the effects of the
paroxyfm being allowed to return, from the neglect of
that medicine.

There are many fymptoms which would forbid the
ufe of Bark, did they occur diftinct and independent
of Intermittent Fever, fuch as Cough, difficulty of

D   breathing

breathing and pain in the fide : they are frequently brought on by the paroxyfm of Intermittent, and only give way to the ufe of the Bark by which that paroxyfm may be prevented. Such fymptoms do not admit of a diftinct and feparate treatment, but are always aggravated by the ufe of evacuants, more efpecially bleeding, the molt probable means of relief in common Depuratory Fevers.

I have fometimes found a complication of Intermittent and Hectic Fever in the fame perfon and could diftinguifh between the paroxyfms of each ; the Bark, while it cured the Intermittent, has even moderated the Hectic ; though Hectic Fever of itfelf, efpecially as a fymptom of *Phthifis Pulmonalis* does not appear to give way at any time under the ufe of the Bark ; I think it probable, therefore, that although in the treatment of intermittents, complicated with other difeafes, our chief attention fhould be firft directed to the cure of the intermittent, yet it may be neceffary to purfue an indication that may have in view the other diforders, not incompatible with the treatment of the Intermittent. In Dropfies, which accompany Intermittents, I have found more benefit from the ufe of Bark, joined to the neutral falts, and other mild diuretics, than from active purgatives, which always tend to protract the Intermittent. Every returning paroxyfm of an ague confirms more and more the caufe of that difeafe which it has produced.

It is not intended to infinuate, that no cafes do occur, in which it may not be prudent to adminifter a vomit, efpecially to perfons fubject to bilious accumulations in the ftomach ; but this is more with a view of removing an obftacle to the operation of Bark, than as neceffary to render it fafe ; and I have frequently feen naufea and vomiting fo much a fymptom of the paroxyfm, that they gave way only to the free ufe of Bark itfelf.

A practice

A practice more abfurd than that of preceding e-
vacuations has been adopted and recommended, *viz.*
That of evacuating by purgatives after the cure had
been compleated by the Bark, this feldom fails of
bringing back the Intermittent, as one caufe of indu-
cing debility, the moft favourable ftate of the body
for the attack of Intermittent Fevers.

In perfons who are rendered coftive by the ufe of
the Bark, I would recommend the common practice
of giving a few grains of Rhubarb, or *Pilul Rufi.* If
on the other hand it fhould prove purgative, a few
drops of the *Thebaic Tincture* is the ufual and beft
means of checking that operation.

Intermittent Fevers are frequently fo very anomalous in their appearances, and affume fo much the
character of other difeafes, that an unfkilful, or inat-
tentive practitioner may be deceived ; they, however,
under any form or any type, generally give way to
the Bark ; fometimes aided and affifted by other
means.

Much experience is required to detect the paroxyfm
difguifed under different forms, and although the Bark
is our beft remedy, yet the violence of fome fymptoms which accompany the difeafe, and interrupt its
natural and ufual form, render it neceffary that other
remedies be occafionally employed, as adapted to the
particular circumftances of the cafe ; not always fo
much with a view of rendering the Bark a fafe remedy, as of rendering it an effectual one.

Such anomalous appearances are greatly influenced
by the nature of the prevailing epidemic of the feafon, they more particularly interrupt the progrefs of
the cold fit, and are generally much aggravated during the paroxyfm of the Intermittent.

The diforders which I have feen complicated with
Intermittent Fevers have been chiefly thofe of a bilious kind, occuring in the autumnal feafon, fuch as
violent and exceffive vomiting, Diarrhæa, with pain in
the

the bowels, Cholera Morbus, periodical Head-Achs, Pain in the Side, and frequent inflammatory disorders, such as Pleurify, Peripneumony, and the acute Rheumatism, and even sometimes spasmodic diseases, terminating in apoplexy and death.

Such diseases have been suppofed to be only varieties of the Intermittent Paroxyfm, and have been believed to give way to the fame remedy which cures an ague.

I think, however, we have no analogy in nature to support this doctrine, and it seems highly unreasonable to suppofe, that the fame caufe can produce fuch a diverfity of appearances. Indeed the influence of the prevailing epidemic diseafes on fporadic complaints evidently fhew, that the human body is fubject to be acted upon by more than one caufe at a time, and that the remedies to be employed fhould have a view to fuch a diverfity of circumftances. It therefore becomes the object of the phyfician to know which of two or more difeafes are moft deferving of his early attention, which the human body may labour under at the fame time. He will generally find, that as the paroxyfm of an Intermittent Fever excites fuch violent action in the fyftem, and generally aggravates the fymptoms of other difeafes, it ought to be early removed.

In fome cafes however I have feen fuch active appearances of inflammation prevailing in a perfon labouring under an Intermittent Fever, and fo much encreafed in the hot fit, that unlefs a quantity of blood had been taken, which was always fizy, the patient moft probably would have died. Such fymptoms of inflammation retard the cure of the Intermittent, and therefore are in the firft inftance to be removed.

This may probably explain the reafon why inflammatory Fevers in their decline often affume the appearance of Intermittent difeafes.

In

In the same manner it may be neceffary to remove Symptomatic Vomiting, Cholera Morbus, and the like, by remedies peculiarly adapted to thefe difeafes, before the ftomach will fo far favour the action of Bark as to enable it to cure the Intermittent Fever.

Peruvian Bark is chiefly adapted to the cure of genuine and idiopathic Intermittent Fevers, and not thofe of a fymptomatic nature, which frequently require remedies of a different kind.

It is impoffible in a treatife of this nature to point out more minutely the circumftances which ought to regulate and direct the conduct of practitioners in the treatment of complicated intermittents, and the condition of the habit, which may render neceffary the previous ufe of other remedies, or the combination of them with Bark, in order to render its operation either fafe or effectual.

It appears to me, that the advantages at any time derived from the ufe of other remedies, depend upon their having removed fome other difeafe, which may have protracted the ague, or interrupted the action of Bark in the cure of the Intermittent, and not on their obviating future effects which have been falfely attributed to Bark, while they are chiefly produced by the obftinacy and imperfect treatment of the Intermittent Fever.

Both the Vernal and Autumnal Intermittents of Dr. Sydenham yield to it, the latter however fometimes with more obftinacy than the former.

In feveral cafes I have experienced the efficacy of the Red Bark in removing Tertians and Quartans which had refifted the common Bark, this however is not to be wondered at, when we confider the diverfity in the power even of different kinds of the common Bark in general ufe.

I hope I fhall be excufed in digreffing fo far, as to mention the ufeful effects I have frequently experienced from the exhibition of Opium in Intermittent Fevers.

vers. We are chiefly indebted for this practice to Dr.
Lind. It moderates fo effectually the force of the pa-
roxyfm, by fhortening the duration of the cold fit, as
well as by diminifhing the violence of the hot fit, that
I had often flattered myfelf, it was capable of curing
Intermittents. In this however I was difappointed.

From the experience of it in many hundred cafes, I
conclude with Dr. Lind, " That an Opiate given
" foon after the commencement of the hot fit, by a-
" bating the violence and leffening the duration of the
" Fever, preferves the conftitution fo entirely unin-
" jured, that fince I ufed Opium in Agues neither a
" Dropfy nor Jaundice has attacked any of my pa-
" tients in thefe difeafes." The manner in which I
employ it, is either by giving a grain of the Thebaic
extract upon the acceffion of the cold fit, or twenty
drops of the Thebaic Tincture upon the acceffion of
the hot fit, the action of the former being later from
its flower folubility.

The Red Bark is fo much warmer than the other,
that it would feem to anfwer all the purpofes derived
from the union of Cordials, Aromatics, Serpentaria,
and the like, fo much recommended in the obftinate
Quartan Intermittents of elderly people.

Some difference in opinion has prevailed regarding
the manner of giving the Bark. Moft practitioners
concur in thinking, that it cures intermittents more
readily when taken in fubftance than in any other form.
In this ftate, it is both a bulky and naufeous dofe in
the quantity neceffary to cure an Intermittent; at any
rate, it ought rather to be diffufed in fome liquid, than
given in the form of an electuary or pills, which are
fometimes difficultly foluble.

I have found milk cover the tafte of Bark, and
make it more acceptable to children than any other
vehicle. The extract of Liquorice diffolved in water, may be likewife employed to cover the tafte of
Bark.

Bark. Its tafte is alfo corrected by wine, efpecially by Old Hock.

It would appear from the general preference given to Bark in fubftance, that its decoction, infufion, or tincture, are found too weak in any quantity for the purpofe of curing Intermittents, otherwife as they are much lighter to the ftomach and act more quickly, they fhould be preferred.

I hope to make it appear, that in this refpect the Red Bark has the advantage of any other kind now in ufe, fince either its infufion or decoction will cure Intermittents, and its powder in a much fmaller dofe than that of common Bark will produce fimilar effects.

The beft time for giving the Bark is in the intermiffion between the paroxyfms and when the ftomach is empty. In Quartan Fevers, where there are two days of Apyrexia, we fhould be particularly defirous of getting down a larger quantity on the day immediately preceding the approaching paroxyfm, and in other Intermittents as near the period of the returning paroxyfm as the ftomach will bear it.

A very prevailing argument in favour of the Red Bark has been fuggefted to me both by apothecaries and their patients, viz. that it will cure when taken in half the quantity which has been found neceffary of other Bark. I am likewife perfuaded from a great variety of trials, that while other Bark only gradually weakens the force of the Intermittent Fever, the Red Peruvian Bark will frequently obviate the return of a fecond paroxyfm. It is feldom I have found it neceffary to give more than half a dram every two hours in the interval of the fit, and in no one Intermittent, even of a Quartan type, have I found it neceffary to give more than fix drams between the paroxyfms. I have frequently known double that quantity of common Bark fail to produce the defired effect.

It is however unneceffary to limit the dofe. One dram may be given every hour, if the ftomach will retain

tain it, and will perhaps in some cases remove the disease more quickly than a smaller quantity given at longer periods *.

The following facts will best determine how far I am justified in favouring the opinion of the superior excellence of the Red Bark.

EDWARD VIRGOE, aged 21, had laboured under an Intermittent Fever five months. It was first a Tertian and afterwards became a regular Quotidian, accompanied with Cough, Dyspnœa, and Hoarseness, particularly in the paroxysm. The common Peruvian Bark, given in the dose of one dram every hour, prevented the return of the paroxysm for a few days; the patient however relapsed notwithstanding the Bark was continued, he was at last cured by taking one dram of the Red Bark every second hour for the space of ten days.

It appears from the above case of Edward Virgoe, and several others which have occurred to me, that Intermittent Fevers, which had resisted common Bark and other remedies, have yielded to the Red Bark, even under very complicated and unfavourable appearances.

In this opinion I am likewise justified by the experience and testimony of many eminent practitioners, and so decided are they in its favour, that the demand for it every day increases, especially in some of the neighbouring counties where Intermittents are not only more frequent, but more obstinate.

I began

---

* I have likewise frequently adopted the practice recommended by Dr. Home, of giving the Bark as soon as the sweating fit of the Fever has sufficiently carried off the hot fit; this is particularly proper when the interval is short between the paroxysms.

I began now to fufpect that its powers were even fufficient in cold infufion in moft cafes to cure Intermittent Fevers, and in all other cafes to anfwer every purpofe which might be expected from common Peruvian Bark, in any form in which it had hitherto been employed.

The following facts are fufficient to authorife this opinion.

JAMES YOUNGMAN, aged fixteen, had laboured under a Tertian Intermittent many months ; it was accompanied with a fevere cough, and his ftrength was confiderably impaired. He was ordered to take four ounces of the cold infufion of the Red Bark every third hour ; after taking it for two days, the paroxyfm did not return. Its ufe was perfevered in fourteen days, and he continued perfectly well.

THOMAS PUGH, aged twenty-two, from Woolwich in Kent, had an Intermittent Fever of five weeks ftanding of the Tertian type, accompanied with a confiderable degree of cough, occurring efpecially in the cold fit ; he took the cold infufion, as recommended in the former cafe, after which time he had only one flight return of the paroxyfm, his cough is likewife totally removed.

JOHN WELDING, aged twenty-five, had an Intermittent Fever at Sheernefs, for near a year, he took large quantities of common Peruvian Bark, without effect. It was of the Quotidian type. After taking a cold infufion of the Red Bark in the quantity of a quart in twenty-four hours for three days, the paroxyfm difappeared and has never fince returned. It proved at firft purgative, but foon loft that effect. His ftrength and appetite were greatly improved under the ufe of the infufion.

E   Since

Since the firſt edition of this work, I have had frequent experience of the efficacy of the cold infuſion
in the cure of Intermittent Fevers, but it requires to
be continued for a greater length of time than is neceſſary, when the Bark is taken in ſubſtance with
wine. I have however ſeen caſes where the ſtomach
had rejected the Bark in ſubſtance, which yielded readily to the cold infuſion when taken in the doſe of
four ounces, every two hours in the interval of the paroxyſm. I have likewiſe met with caſes of Intermittent Fevers ſo complicated with other diſorders, as to
make it neceſſary to act more gradually and cautiouſ-
ly on the conſtitution, by the uſe of the cold infuſion,
than by giving the Bark in ſubſtance in the doſes neceſſary for inſtantly removing the Intermittent Fever.

As I conſider a perſeverance in the uſe of the Bark
proper for ſome time after the paroxyſm has been removed, I think that it is only neceſſary to have recourſe to the cold infuſion for that purpoſe, and it will
be found a much more agreeable preparation than any
other. I have found great benefit from continuing its
uſe in the convaleſcent ſtate of perſons after Intermittent and other Fevers.

I could here enumerate a great variety of caſes
which have occurred to me, both in public and private practice, in confirmation of the general doctrines
I have now laid down, but I think it better to mention the general reſult of a ſucceſsful practice.

The cold infuſion employed in the above caſes was
prepared by pouring a quart of cold water on two
ounces of the Red Bark in fine powder, frequently
agitating them for the ſpace of twenty-four hours *.

OF

* Though the caſes above mentioned, evidently prove that the
cold infuſion will cure Agues, yet they are not intended to divert the
attention from other more effectual means of giving this medicine.

OF ITS USE IN THE CURE OF OTHER FEVERS.

Remittent Fevers are frequently equally steady in their periods of remission and accession with those of the intermittent kind, but the *Apyrexia* being less perfect and complete, has given rise to many doubts regarding the safety of giving Bark. As remittent Fevers are more particularly marked by appearances indicating the prevalence of bile in the stomach ; the propriety of giving an emetic, prior to the use of the Bark, seems well founded, and the remission is frequently rendered more complete by such a practice.

In the remitting fevers, however, of warm climates, the accession of the paroxysm is so extremely violent, and the strength of the patient so quickly exhausted, that it becomes absolutely necessary to catch the first opportunity of the most trifling remission, and to give the Bark with the same freedom that you would do in common Intermittents.

In Remitting Fevers, the concomitant symptoms are more to be regarded than in Intermittents, because they more especially disturb, interrupt and shorten the periods of remission ; even in the warmest climates, and under the severest prejudices, it has been found necessary to take away a few ounces of blood to relieve pain in the head, oppression in breathing, an intense dry heat on the skin, and other symptoms protracting the paroxysm ; by such means the remission has been brought on, and the Bark given with greater effect.

The Fevers of this country seldom have regular remissions, until they have been properly treated by evacuations ; the inflammatory by bleeding, and the bilious by vomiting and purging.

When Fevers are brought into a state of obvious remission ; that is, when the pulse becomes from ten to twenty flower at fome particular time in the twenty-four hours; when the reftlefsnefs, anxiety, and tendency to delirium abate; when the mouth and fauces are moift; when the organs of fecretion, and efpecially the fkin are more.open and pervious, fuch fymptoms of remiffion admit the ufe of Bark with the fame freedom as in Intermitting Fevers.

. The Acute Rheumatifm, notwithftanding its in-flammatory attack, and the appearance of the blood, and though the joints often continue inflamed feveral weeks, very• early affumes the form of a Remitting Fever.

. Under fuch circumftances, a perfeverance in the Antiphlogiftic plan is generally found to be ineffectual. I have in feveral cafes of this kind, employed a cold infufion of the Red Bark, and the difeafe feemed to give way only to this treatment.

The Acute Rheumatifm in its remiffions affumes the form of a double Tertian, and the patient is frequently.greatly exhaufted by the profufe fweatings which terminate the paroxyfm.

It is in fuch cafes that I would particularly recommend the ufe of Bark. I have found this practice more fuccefsful, and it muft be allowed to be more rational, than the ufe of *Volatiles* and *Guaiac*.

The tedioufnefs, as well as the inefficacy of the antiphlogiftic practice in the cure of the Acute Rheumatifm, has frequently directed my attention to that difeafe in a particular manner, and after being repeatedly difappointed and diffatisfied by purfuing the ufual mode of treatment recommended by the beft writers, either antient or modern, on the fubject, I was determined to adopt other means, which I think have proved more fuccefsful. The Rheumatic Fever appears to me, notwithftanding the violence of its inflammatory fymptoms, to be an Intermittent Fever in a ftate of difguife, and its periods are evidently, as I have already mentioned, thofe of a double *Tertian*.

Its

Its inflammatory symptoms, however, ought in a certain degree to be reduced by moderate bleeding, occafional purging, and great dilution, before it can be treated as an Intermittent Fever; it does not however appear to me incompatible with the ideas of its being inflammatory, to have recourfe to the moderate ufe of the Bark, to obviate the weaknefs which may be induced by the neceffary ufe of the lancet, nor does it appear repugnant to the idea of its being intermittent, that the inflammatory fymptoms which render the intermittent anomolous and irregular, fhould be moderated and checked by an antiphlogiftic treatment accompanying the ufe of Bark. I have found in many cafes by this practice, the Rheumatic Fever greatly fhortened, and the debility and torpor in the joints, which is frequently the effect of that difeafe, together with the difpofition to the Chronic Rheumatifm, generally prevented. The Acute Rheumatifm appears from the ftate of the pulfe, the tendency to profufe fweating, the depofition in the urine, the frequent acceffion of chilly paroxyfm to be an Intermittent Fever, which is probably prevented from affuming the more ufual and natural form of that difeafe by the inflammatory action on the joints, which I have fometimes feen merely local, (i. e.) unaccompanied with any general Inflammatory Fever in the habit : in fuch cafes, while leeches have been applied to the joints, and the hemorrhage from them encouraged by fomentations, I have given Bark freely, which I have never known to increafe the inflammatory fymptoms while the ufual means of promoting inflammation were guarded againft. General and vague maxims, applied in reafoning on the action of Bark, and its tendency in all cafes to promote inflammation are therefore ill founded, and had they not been corrected by experience and obfervation, would have deprived us the advantages we have derived from the ufe of this medicine in the cure of Rheumatic Fevers, Scarophulous Inflammation,

flammation, and perhaps a great variety of other difeafes. I have feen in delicate and irritable habits
Rheumatic Inflammation on the joints, accompanied
with a low Nervous Fever, which gave way only to
Bark and Sedatives.

In the Rheumatic Fever I generally begin about
the feventh day from the attack with the cold infufion of the Red Bark, in the dofe of three ounces every
two or three hours, until the evening paroxyfm comes
on ; nor am I, by this practice, in any degree, divert-
ed either from, general or local bleeding, or evacuations by ftool, when the circumftances occur which
may render them neceffary.

In that *low Fever*, which Huxham has fo well defcribed under the title of nervous, I have prefcribed
the cold infufion of the Red Bark with advantage,
where the fkin hath been foft, and the pulfe under one
hundred and ten.

In fuch Fevers, it chiefly acts as a Cordial in fup-
porting the *Vis Vitæ*, and for which reafon I think it
may be given with advantage in the decline of all Fevers, even where the fymptoms on the attack of the
difeafe were evidently inflammatory.

Almoft every Fever remits in its decline.

Fevers originally putrid and malignant, as arifing
from *Miafmata* and putrid vapour, very feldom occur
in this City, they are moft generally to be found in
Fleets and in Camps, and in fituations where air ftag-
nates or where animal bodies are confined in a clofe
place.

In fuch Fevers, the cure is effected by vomiting and
warm Cordials ; of the laft is the Red Bark in an
eminent degree.

It may be infufed in wine, which will render its
operation more antifeptic.

It intimately unites with the feveral acids, from
which a very favourable operation in fuch cafes may
be expected.

In

In the Putrid Fever, attended with a gangrenous Sore Throat, I have in many inſtances experienced the efficacy of Bark, but care ſhould be taken not to confound this diſeaſe with the *Angina Mucoſa* of Dr. Huxham, or the *Angina Eryſipelatoſa* of Dr. Grant, diſeaſes, though contagious, and accompanied with eruptions on the ſkin, evidently of the moſt inflammatory nature, and requiring the uſe of evacuations.

In *Petechial Fevers*, with ſymptoms of great proſtration of ſtrength, I have frequently ſeen the pulſe not much quickened, and the animal heat very little encreaſed. In ſuch caſes I would recommend the uſe of the Red Bark infuſed in Old Hock.

I have ſeen a *Jail Fever* with no other diſtinguiſhing ſymptoms than Petechiæ and Debility; the tongue clean, the pulſe moderate though ſmall, and the ſecretions apparently not diſturbed. In that caſe, bliſters on the extremities, Bark and Wine, are the beſt remedies.

In general, we find that Fevers marked with ſymptoms of debility are chiefly found to remit, and therefore they admit of the uſe of Bark ; by increaſing the tone and vigor of the ſyſtem, it oppoſes a returning paroxyſm.

Fevers of more violent action, ſuch as we denominate inflammatory, do not remit at leaſt ſo obviouſly, until that action is moderated by Evacuations, ſo that they chiefly remit only in their decline.

The primary attack of moſt Fevers in this country, is attended with violent action, which is beſt moderated ſometimes by the prudent uſe of the lancet, but moſt frequently, by the Tart. Emetic, with the infuſion of Sena and ſome of the neutral ſalts. All Fevers beginning with a violent Rigor, and followed with great heat, require early evacuations, while ſuch as creep on ſlowly and imperceptibly in the beginning, with giddineſs in the head, rather than acute pain, much anxiety and watchfulneſs, tremor and debility,

give

give way to Opiates, Bark, Serpentaria and Wine? Blisters applied to the arms are extremely useful in keeping up the pulse, especially if the skin be soft, the tongue and fauces sufficiently moist; much more may be learned by attending to this last appearance than is generally known.

It is necessary, however, to distinguish between the dryness of the tongue and mouth, as a symptom of the Fever, and as arising from the circumstance of a patient sleeping with his mouth open.

I hope the observations here offered to the public, on the subject of this invaluable remedy, will remove all prejudices against a Peruvian Bark of a large and coarser appearance, than is generally employed.

It is at present in very great demand, the difficulty of procuring it will not, I hope, instigate Druggists and Dealers in the article, to substitute at any time a spurious kind in its room.

Extract of a Letter from Mr. EDWARD JACOB, *jun. an eminent Surgeon, at Feversham, in Kent.*

I HAVE had such repeated opportunities of trying the Red Bark, that I hope to be able to give you every satisfaction you can wish, of its superior efficacy over the Bark in common use.

Our situation being in a country not far distant from the marshes, renders the inhabitants more exposed to intermittent complaints than those of cities and more inland counties; and I assure you, before the use of the Red Bark was known, the Ague, from its particular stubbornness (as we thought, or what we have since observed from the want of efficacy in the other Bark) might be truly called the *opprobrium Medicorum*; but now I think that stain is entirely removed, for I have not met with one single case, where (when I could per-
suade

suade the patient to adhere steadily to my advice) I have ever found the Red Bark to fail.

The Peruvian Bark, with which I was formerly provided, was of the best kind, and always had in the quill; yet, it was even much inferior in its effects to what ought reasonably to be expected from it.

The first knowledge that ever we had of the Red Bark was in March, 1781, when a few pounds were sent us to try its effects; our Druggist informed us, that it was then in use at St. Bartholomew's Hospital, and was there found effectual. It remained in our house till May following, when I was attacked with an Ague; I did not at first think of trying the new Bark, for I guessed, by taking of the old in great quantity, which would not be disagreeable to me, that my Ague would soon leave me; but, to my great disappointment, fit succeeded fit, without shewing the least good effect of the remedy applied. I was then determined to try the new Bark; but finding my Ague stubborn, I emptied the *Primæ Viæ* by an emetic and carthatic, and immediately, on the fever going off, I took one dram of the Red Bark in fine powder, mixed with three ounces of the decoction, and a small quantity of the tincture, which being a draught well loaded with Bark, at first seemed to sit uneasy on the stomach (which I have several times on being first taken, found the case with some of my patients, but which never took from its effect) but, after resting for some time I found myself in a state to continue its use. My Ague from that time kept off; till, from omitting the Bark too soon, and finding myself quite well, in a few weeks after, flight symptoms appeared again; but which never formed a direct fit; the Bark being again repeated, eradicated the disorder.

My father, now in the 70th year of his age, has since that time been attacked with an Ague; but, from his having been before seized with a chilliness

F   without

without fubfequent fever, fuffered himfelf to have three fits without trying any remedy ; being then convinced of the reality of the Ague, applied to this Bark in decoction, with fome tincture made of the fame ; he found it fo effectual, that after taking it, the fit did not once return ; he omitted the Bark too foon, and had one flight fit ; but, after repeating the fame remedy a few days, it has not fince returned, and he is now perfectly well.

The patients who have received immediate relief from the new Bark, are fo numerous, that I cannot, with any degree of certainty, guefs the number.

The quantity that we have ufed from July 1781, to the prefent time, is upwards of fixty pounds. The manner in which we have ufed it, with a view to prevent the return of an approaching paroxyfm, is by giving half a dram or one dram of the powder, mixed with two ounces of the decoction, and a fmall quantity of the tincture. The decoction we have ufed, has been prepared by boiling three ounces of the grofs powder boiled in two quarts of water to one quart.

When the patient has been of a more delicate frame, or when it has been ufed as a corroborant, we have given the decoction and tincture without the powder. When the Ague has been but recently contracted, we have feldom trufted to any thing but the Bark, but when ftubborn, evacuants have been firft given ; half an ounce of the powder has very frequently cured an Ague without evacuants, or more Bark, when the fit was a recent one ; we have now entirely left off giving the old Bark in any form fince we have found fo good effects from the Red Bark, and we ufed the refin of Red Bark with good fuccefs in many cafes, where the patient could take it only in the form of pills. I heartily hope the above account will prove fatisfactory to you, if not I fhall think myfelf very

happy

happy in anſwering at any time any future enquiries.

I am, Sir,

with great reſpeċt,

your moſt obedient

and humble ſervant,

EDWARD JACOB, Jun.

Feverſham, June 23, 1782.

A Letter from Mr. Boys, *an eminent Surgeon aud Apothecary at Sandwich, in Kent.*

SIR,

I AM ſorry it is not in my power to ſend you particular caſes of the effeċts of the Red Bark : having
made no memorandums, I cah only ſay, in general,
that it is a much better kind of Bark, than any I have
been able to procure hitherto. Whether the Intermittents have been worſe than common, or the Grey
Bark has been of inferior quality, I will not take upon me to determine ; but we were unuſually foiled in
our attempts to cure that complaint, till we were furniſhed with the new Bark.

I can have no inducement to give a higher character of this drug, than it deſerves : but I do aſſure you,
that ſince I began to uſe it, it has not once failed me,
when the patient has taken it in a proper manner.
Hence, I have a firm perſuaſion of its ſuperior efficacy, and I am the more confirmed in my opinion, by
knowing that my ſentiments correſpond with yours,
and with thoſe of all my medical acquaintance, in this
neighbourhood, who have made trial of it. My beſt
wiſhes attend your publication ; not only becauſe I
am perſuaded it will operate to the advantage of mankind, by extending the knowledge and uſe of this valuable medicine, but likewiſe, becauſe I am in hopes

it

it will haften a frefh importation of the article, of
which I have very little left, and I know not where
to go for a fupply.

I have the honour to be,

Sir,

Your moft obedient fervant,

W. B O Y S.

Sandwich, June 19, 1782.

In addition to the character given of the Red Bark, by fe-
veral Practitioners in the country, is a Letter I re-
ceived a few days ago from Sir WILLIAM BISHOP, *an*
eminent Surgeon, at Maidftone, in Kent, dated June
16*th.*

I N which he fhews, by a variety of cafes, that in
the cure of Intermittents, in that part of the world,
the Red Bark had not only been infinitely more effectual than the common Peruvian Bark or any other remedy ufually employed, but that it had radically cured where Cold Bathing, Emetics, Opiates, Bitters, and
Friction failed, and where the beft Pale Bark, both by
itfelf, and with a variety of other medicines, was exhibited without effect, even to the quantity of fixteen
ounces. The form he gives it in, is that of an electuary, with the addition of a fmall quantity of the effential Oil of Pepper-mint and Carraway-feed ; and
he feldom has occafion, he fays, to ufe more than from
four to fix ounces. One cafe he mentions, where a
Sphacelus had taken place in a Malignant Fever, accompanied with delirium and every other bad fymptom, the patient was recovered by the ufe of the Red
Bark, Anodynes, and Wine ; the Sphacelus feparating kindly, the Fever and every bad fymptom fpeedi-
ly gave way.

I am

I am likewise favoured with another Letter from Sir WILLIAM BISHOP, *dated June 23, which contains many senfible and judicious hints upon this subject.*

H E recommends the grinding Red Bark into the fineft powder by a mill, and afterwards fifting it through the fineft cyprefs fieve, in order that it may fit lightly on the ftomach, and that its parts may be uniformly blended together : he obferves, that if, af-ter breaking the Bark, you examine the broken pieces, by means of a glafs in the fun-fhine, you will fee the refin like fpangles of gold between the internal woody fibres and the outward grey coat ; by comparing this appearance with the beft Pale Bark, you will difcover the Red Bark contains a much larger quantity of refinous parts.

Two fcruples of Bark, as coarfely powdered as is commonly fold by the apothecaries, will be as difagreeable to take, as a dram or four fcruples finely ground. One dram is the dofe he generally gave to an adult, although he fometimes met with patients, who would choofe to take two or three drams at a dofe, and thofe dofes, at longer intervals ; and from experience he found that, when this laft practice agreed with the ftomach, it was the moft effectual way of curing the Intermittent. One patient took three ounces in twelve hours, and had no return of an obftinate *Quartan*.

Many Intermittents, which could not be cured by fmall dofes, were removed by larger dofes, more frequently employed.

He mentions that a period of five or fix hours, before the expected return of a *Quotidian* or *Tertian* Intermittent, is all that is neceffary for the taking a fufficient quantity of the Red Bark, in order to obviate the approaching paroxyfm. Some have been cured by taking two ounces, as quick as poffible after the
fit,

fit, but such as persevered in the use of it, until four or six ounces were taken, acted most prudently, their health was soonest and most perfectly restored. He likewise observes, that such persons ought to take the greatest quantity of Bark, whose blood is in the most dissolved state, and where the fibres have been relaxed by the most profuse sweating, which frequently takes place in obstinate *Quartans*.

Quotidians·require but a small quantity, *Tertians* will yield to fewer doses than our *Kentish Quartans*.

He found it better to get down five or six ounces of Bark, in eight or ten days, than to allow a larger time for the same quantity. Labouring people, who work out of doors in cold foggy mornings, either in marshy grounds or wet lands, from sun-rise till the dew of the evening, and who sweat profusely, and suffer their wet linen to dry on their backs, will not be safe from a return of the *Quartan* Fever, with less than five ounces taken after the last fit, as the good effects of all they had taken before, are very probably carried off and dissipated in the ensuing paroxysm.

He likewise observes, that the Bark in large doses will not cure the irregular fits of an ague and fever, which so often afflict persons, when matter is forming in the lungs, in the *Phthisis Pulmonalis*.

The letters from Sir William Bishop, which convey these observations, likewise contain much valuable information on other medical subjects, and I hope the ingenious author will, on some future occasion, favour the public with them.

A Letter from Dr. Withering, an eminent Physician, at Birmingham, June 29, 1782.

DEAR SIR,

I AM much pleased that you have undertaken to give us an account of the Red Peruvian Bark. A publication upon that subject, cannot fail to produce good effects, by removing the prejudices of some, the ignorance of others, and ultimately by exciting our merchants to obtain liberal supplies of a medicine so truly valuable. The universal prevalence of Intermittent Fevers this spring gave us a large experience of its effects. We have not many Intermittents in the town of Birmingham ; but in other parts of the county of Warwick, in Staffordshire, Shropshire, Worcestershire, and Oxfordshire, so far as my rides extend, they have been more general than ever known before in the memory of the oldest practitioners I have conversed with.

After taking pains to recommend the Red Bark to all the Apothecaries I met with, and consequently not less anxious to learn their observations relative to its effects, I can say, that they unanimously concur in asserting, " that they have never been disappointed in " their expectations, when they exhibited it to patients " labouring under Intermittents."

Now the Intermittents which I have seen, have pretty generally assumed the *Tertian* Type in light soils, and the *Quartan* Type in clayey countries. Of the latter, I have the care of some which were from six to eighteen months duration, originating in Kent and Essex. I have known two instances only of *Quartans* in which the Red Bark did not prevent a recurrence of the Fits ; I saw one of these patients afterwards ; he laboured under hepatic and anasarcous symptoms, these were removed in a fortnight by the usual

uſual methods, and then the Bark affected a cure. The other caſe was at a conſiderable diſtance from this place, and I have not yet learnt its termination.

It will require ſome farther experience to aſcertain the neceſſary doſes. I know ſome practitioners who have given one or two drams every four hours betwixt the fits, but I have never had occaſion to give more than thirty or forty grains at ſimilar intervals of time.

As to its preparations I can ſay but little, preferring always in my own practice the ſimple powder; but, I am told, that it makes a very rich tincture, and I have ſeen decoctions of it very high coloured and turbid.

But it may be aſked—Would not the common or Quill Bark, have produced ſimilar good effects in the Intermittents of the preſent year? From the reports of other practitioners, I believe it would not. From my own experience, I can give no other anſwer to ſuch a queſtion, than, by ſaying, that after frequent and almoſt continual diſappointments, from the uſe of common Quill Bark, I have not tried to cure a ſingle Intermittent with it for more than ſeven years paſt; relying entirely upon the uſe of evacuants, opium, and metallic ſalts. To render this laſt aſſertion reconcilable to the daily experience of others, it may be neceſſary to obſerve, that with us, a Phyſician is ſeldom conſulted in Fevers of the intermittent kind, unleſs ſome unuſual appearance, or ſome uncommon obſtinacy in the recurrence of the attack, alarms the patient or his friends.

Whether the medicine in queſtion be the product of the ſame tree from which the Quill Bark is derived, may be difficult to determine, but I am perſuaded it is the medicine that was uſed by Morton and Sydenham, or its efficacy could never have been ſo proverbial. I know not what could firſt induce the writers upon the Materia Medica, to prefer the Quill Bark, but I know if you were to aſk an Engliſh Tanner,

whether

whether the Bark from the trunk, or that from the twigs of the Oak is the strongest, he would laugh at your ignorance.

I communicated the contents of your letter to my worthy friend and colleague Dr. Ash, together with my opinion upon the subject; he authorises me to say, that the result of his experience perfectly coincides with mine.

I remain,
with the truest esteem,
Yours',
W. WITHERING.

P. S. Thus, my good friend, have I freely communicated my sentiments and observations upon the subject you have in hand.

You are at liberty to make what use you please of them, and I shall think myself happy in having contributed a little towards the extensive usefulness of your design. I have seen no bad effects from it, notwithstanding the reports of some of the London Druggists, but their motives were too evident to need a comment.

I feel, with you, that distance alone has interrupted our communications; but I feel too, that distance can never abate the regard with which I once more subscribe myself,

Affectionately,
Your's,

W. W.

To Dr. Saunders, Jefferies Square.

A Letter from Mr. SHERWIN, an ingenious Surgeon at Enfield.

SIR,

SINCE we have used the Red Peruvian Bark, we have had great success in curing Agues and Intermittent Fevers. These disorders returned so frequent-

ly,

ly after the ufe of the beft Peruvian Bark, which we could formerly procure, that our credit, as well as that of the medicine, began to fail very much, and numbers of our patients got into the hands of perfons unacquainted with phyfic. They feemed willing to truft rather to noftrums and charms, than to a medicine which they have feen fo repeatedly fail, and to the effects of which they very ingenioufly afcribe every ache and pain which continue after Agues, or which are the confequences of Agues when imperfectly cured.

I could give a very ample detail of the inefficacy of the common Bark ; but as the complaint is general, it would be unneceffary. I fhall, therefore, only take the liberty to prefent you with the more agreeable hiftory of a few cafes that have immediately yielded to the Red Peruvian Bark, and that even when given in very moderate dofes, after having obftinately refifted large quantities of the other.

RICHARD PARSLEY, a hard working young man, about twenty-eight years of age, was feized with an Ague laft autumn, which continued either as a *Quotidian, Tertian* or *Quartan,* with very little intermiffion, till the 20th of April laft, when I faw him accidentally. He gave me the following account :

That he had taken every thing that had been recommended to him ; and that his mafter, a worthy and refpectable gentleman in this neighbourhood, had procured him the advice of different practitioners of phyfic, who had prefcribed Bark for him in large quantities, but without benefit, as the diforder always returned with greater violence after it had been checked a few days. He added, that the laft medicine which he had been advifed to take, was half an ounce of Allum boiled in a pint of Ale, to half a pint, which he was ordered to repeat three different times, as foon as he felt the Ague approach. He had in

this

this way, he faid, taken feven half ounces upon the approach of as many different fits of the Ague.

I fufpected, that fo large a quantity of Allum, taken at one dofe, muft have produced fome dangerous effects, and doubted whether he had fwallowed the whole of it; but I found, upon more particular enquiry, that he really fwallowed half an ounce each time, as he fays; that he ftirred it up from the bottom and felt it gritty in his teeth. It gave him great pain in his ftomach. So large a quantity of Allum being taken at one dofe, without materially injuring the fyftem, is a fact which I am perfuaded you will think worthy of notice.

I took compaffion upon this poor fellow, and fent him eight papers of the Red Peruvian Bark, containing a dram in each, which he finifhed in two intermediate days, the diforder being then a *Quartan.* It is now feven weeks fince he took this medicine, and he has had no return of the complaint, but, to ufe an expreffion of his own, " hath felt ever fince as if he had a new infide."

ANNE PIGOT, a poor girl, fourteen years old, has been afflicted with an Ague fince Autumn laft in the fame family with Parfley, and has alfo tried various methods without effect. I prefcribed fmall dofes of the *Vitriolum Cærul.* fourteen days without gaining any advantage. I gave her alfo three or four dofes of a hot aromatic powder, confifting of Bay Berries and Caian Pepper, which I have frequently feen to cure, but fhe found no benefit from it. I cannot fay whether this girl had given the common Bark a fair trial. She begged to have fome of the fame medicine that cured her relation, and about a month fince had three papers, containing only one fcruple in each, and I affure you, fhe has been perfectly free from the complaint ever fince.

Mr.

Mr. BARNES, a very ſtrong, hard working man, was ſeized laſt Autumn with a very violent Intermittent Fever, attended with delirium, and apparently much danger. After three or four paroxyſms, I removed the diſorder by giving him one dram of the common Peruvian Bark every hour, and continuing the ſame quantity twice a day afterwards ; but at the end of fourteen days, the Fever returned with the ſame violence as before, and was again ſtopped by the ſame quantity of Bark. From Autumn, till the beginning of laſt April, it returned in this manner (at the end of about fourteen days from the time it was ſtopped) five or ſix times, and always with uncommon violence. The common Peruvian Bark, newly powdered, was repeatedly given.

In the beginning of April, I preſcribed ſix doſes of the Red Peruvian Bark, of one dram each, which he, took upon the going off of the paroxyſm, and I have the pleaſure to inform you, that he has had no return ſince that time ; though I no longer uſed the precaution of continuing the daily uſe of the ſame medicine.

It is not worth while to mention common caſes, where the Red Bark hath cured without the other having had the trial. A few have occurred, and I have not yet had any inſtance of a relapſe.

I was lately called to a very ſevere caſe, like that of Mr. Barnes, where the intermiſſion was not longer than ſix or eight hours. I preſcribed, and the patient took ſix drams of the Red Bark, without interrupting the paroxyſm, which came at the expected hour ; and during, the ſickneſs and horripilatio the Bark ſeemed to be entirely rejected by vomiting ; however, the ſucceeding paroxyſm abated ſomething of its fury. My patient was extremely averſe to Bark in every ſhape ; but my deceiving him, during the following intermiſſion, I got him to ſwallow near the quantity of two drams at once, which rendered him ſo entirely averſe to the medicine, that he would take no more afterwards.

wards. However, to his great joy and my furprize, the Fever left him. He afcribes his cure to three or four draughts of Camphorated Julep, which were preſcribed as a *placebo* upon his refuſing to perſevere in the uſe of the Bàrk. But it may with more juſtice be afcribed to the two drams of Red Bark, taken at one doſe; as I think the firſt fix were in a great meaſure loſt.

I confider the Red Peruvian Bark as a valuable acquiſition to the *Materia Medica*: or, perhaps, it may be only a reſtoration of what was uſed in the days of Sydenham, and ſome time after, when it was common for medical writers to ſay, that this, or that medicine would cure, with as much certainty as Bark would an Ague. An Eulogium to which the Bark in this country has not been of late years entitled.

Before I became acquainted with the ſuperior efficacy of the Red Peruvian Bark, I had tried a variety of Tonics, with very indifferent ſuccefs in general, but now and then with very happy effects, ſuch as the Cuprum Ammoniacum, Vitriolum Cœruleum, Sal Vitrioli, Sal Martis, &c.

WILLIAM KING, an athletic young man, applied to me laſt Autumn, on account of a tedious irregular Intermittent, complicated with acute fixed pain in the fide, and a full ſtrong pulſe, much cough and hoarſeneſs. Under theſe circumſtances, I thought it imprudent to adminiſter the Bark in any form, and had recourſe to a plentiful bleeding, applying bliſters to the part affected, and giving the Decoct. Taraxac, with Tart. Solubile, and honey in large quantities. The diſorder was clearly an Ague; but attended with ſymptoms that threatened a pulmonary confumption. His blood was remarkably ſizy, which induced me to make uſe of *venefection*, four times, which, with the above medicines, and a very ſtrict *antiphlogiſtic* regimen, removed the cough and the pain in the fide,

though

though the Ague ſtill continued. I ventured at laſt to give him eight doſes of common Bark, one dram in each, which removed the Ague ſeveral weeks, and mended his habit much. A continuance of the Bark, twice a day, for ſome time after the Ague ſtopped, brought on a frequent bleeding at the noſe.

The Ague returned twice during the winter, and gave way ſoon to the ſame Bark, and he kept tolerably well till the ſetting-in of cold north-eaſt winds in ſpring, when his Ague again returned, and brought with it the pain in the ſide, a bloated icteric countenance, much hoarſeneſs and cough. After one more bleeding, and the uſe of the Apozem for a month, I gave him ſix drams of the Red Peruvian Bark in April laſt, and have had the pleaſure to ſee him continue free from the Ague ſince that time, though he never repeated the doſe after it ſtopped.

I was unwilling to perſiſt in the uſe of the Bark after the Ague ſtopped, leſt it ſhould increaſe the circulation too much, and again excite a bleeding at the noſe.

I have now indeed almoſt entirely laid aſide the cuſtom of perſevering in the uſe of the Bark, after the Intermittent is ſtopped. I have for ſome time ſuſpected that it anſwers no good purpoſe, and that it may poſſibly now and then be the reaſon why large quantities of good Bark have been given in vain.

It is a well known fact, that every ſpecies of intermittent complaint frequently returns in fourteen days after being interrupted by means of Bark, notwithſtanding that medicine is continued every day. May not the daily uſe of Bark ſo habituate the conſtitution to its effect, as to render it uſeleſs when the Intermittent returns?

I am now ſatisfied when I have put a ſtop to the expected paroxyſm, and preſcribe an anodyne to be in readineſs if ever the cold fit returns, adviſing my patient to have recourſe to the ſame quantity of Bark

which

which he found neceſſary at firſt, and to take it as ſoon as the ſucceeding Fever abates. I adopted this practice a little time before I became acquainted with the ſuperior efficacy of the Red Bark, and thought it an improvement, but ſince that time I have ſucceeded ſo well by giving ſix or eight drams between the paroxyſms, that I have ſeldom had occaſion to repeat a ſingle doſe.

I am,

Sir,

with great reſpeꞔt,

your obliged and

moſt obedient Servant,

JOHN SHERWIN.

Enfield, June 23, 1782.

A Letter from Dr. FOTHERGILL, *an eminent Phyſician, in Harpur-ſtreet.*

DEAR SIR,

IN anſwer to your obliging requeſt, I ſhall now proceed to lay before you the beſt information I can concerning the New Bark, lately introduced into practice, under the name of Cortex Ruber, or Red Bark. As it was found on board a Spaniſh prize, intermixed with a ſmall quantity of common Peruvian Bark, it would ſeem to be no other than the produce of the trunk or large branches of the ſame tree. It ſeems to be poſſeſſed of the ſame ſenſible qualities, only in a much higher degree, hence it yields a much larger proportion of reſinous extraꞔt, and gives a more ſaturated tincture and decoꞔtion than the common Bark. Hence too I have found (according to what you lately obſerved) that the decoꞔtion may be expoſed a long time to the open air, without contracting the degree
of

of acidity which manifeftly takes place in that of the
common Bark in a few days.

. It has been remarked for many years paft, that the
Peruvian Bark has often difappointed the expectation
of the public ; having fallen greatly fhort of that tran-
fcendent degree of efficacy, for which writers of the
laft century have fo highly extolled it. Few Phyficians of extenfive practice, but muft fometimes have
had the mortification to obferve their favourite fpecific entirely baffled by a regular Intermittent without being able to affign any probable caufe for the
defeat, except the ungenuinefs of the medicine ;
whence it would feem to follow, either that the Intermittents of late years have been more obftinate in their
nature, or that the Bark has been of an inferior quality to that which was ufed by Dr. Sydenham, and many of his refpectable cotemporaries, of whofe veracity
we can entertain no reafonable doubt. The latter appears to me to be the more probable, efpecially when
it is confidered that of late years, it has been cuftoma-
ry (for what reafon I cannot imagine) to felect the
Bark of the fmall branches, under the denomination
of Quill Bark, in preference to that of the trunks.
Should this preference hereafter be difcovered to have
been groundlefs (agreeable to what I have long fufpected) it may perhaps contribute to explain the principal caufe of our difappointments.

In the art of tanning, experience has long determined in favour of large Oak Bark, as being greatly preferable to that of the fmaller twigs. Why the reverfe
of this fhould take place in a medicinal view, is by no
means evident, efpecially if the virtue of the Peruvian
Bark keeps pace in any degree with its aftringency.

Spanifh practitioners, as I have been lately informed, are fo well convinced of this, that they always prefer the large Peruvian Bark ; feparating it from the
fmaller fort, for all important purpofes, which the late
capture

capture indeed feems to render probable *. Future obfervations may probably difcover that this was the real genuine Bark, with which our anceftors cured all the various kinds of Intermittents, with a degree of certainty, which now aftonifhes their fucceffors.

If Bark taken from the large branches fhould be found to be more efficacious than that of the fmall, is it not reafonable to believe, that, that of the trunk or rather of the root, might furpafs both in virtue ? Political reafons, however, refpecting the prefervation of the trees, which produce fo important an article of commerce, will, it is to be apprehended, effectually deter the inhabitants from every experiment of this nature.

A very eminent Druggift, who purchafed a large fhare of the above cargo, affures me, that many of the Apothecaries whom he has fupplied with the Red Bark, and who have carefully compared its effects with thofe of the common cortex, make no fcruple of deciding in its favour. He further adds, that the demand for it has increafed fo rapidly of late, that the ftock in hand is already nearly exhaufted.

It is to be regretted, that the paffion for interlarding this fimple febrifuge with other bitters and aftringents (after the ufual complex mode) fhould ftill too much prevail in practice. Under this falfe idea of adding to its efficacy, its virtue may often be greatly diminifhed, and at all events, the refult of the trial muft be rendered extremely equivocal. In juftice to the remedy, and to obviate this uncertainty concerning its effects, I have embraced every opportunity (which fo fhort a fpace would admit) of adminiftring it in its fimple ftate, without intermixing it with other medicines.

H

* This muft not however be confounded with thofe coarfe woody flakes, which conftitute a large proportion of the worft kind of Bark, now in ufe. For thefe, having been already ftripped of the outer rind, together with the refinous cells confift of a mere ligneous fubftance, divefted of medical virtue.

medicines. The cafes in which I have chiefly tried it, have been low putrid Fevers, attended with extreme proftration of ftrength, delirium during the evening exacerbations, with fhort and obfcure remiffions in the day time.

For the fake of brevity I will but mention one inftance. William Henton, a Silk Weaver, laboured under a Fever of this kind, accompanied with Petechiæ, and a profufe nafal Hemorrhage, which laft fymptom generally returned with the exacerbations. To thefe were added cold fweats, muttering delirium, involuntary twitchings, &c. Before I faw him, common Peruvian Bark had been prefcribed, and his cafe pronounced defperate. In this very critical fituation, however, I was not deterred from recommending the Red Bark, in proper diluents, acidulated with Spiritus Vitrioli. Accordingly he took from two fcruples to a dram every two hours, except during the midnight exacerbation, when it was ordered to be cautioufly avoided ; becaufe I have long been convinced by experience, and attentive obfervation, that the febrile fymptoms are generally exafperated when a Bark remedy is given at that period. The medicine agreed, the Hemorrhage abated, and Fever foon fubfided. Some weeks have now elapfed, and I have the fatisfaction to add that lately, on entering his room, he affured me with a joyful countenance, that fince he had taken the Red Bark (or as might now, perhaps, with more propriety be faid, the Real Bark) he had fuffered no fymptoms of relapfe, and that he remained free from all complaints, except a little weaknefs of fight, to which he had been formerly liable. Should this medicine be hereafter found to anfwer as compleatly in the other various types of Fevers, and that in fmaller dofes than the common Bark, it will become highly interelting to the patients, and no lefs pleafing to the Phyfician, efpecially when he fhall be enabled beforehand,

hand, thus to pronounce with real confidence concerning the event.

" *Hi motus, atque hæc certamina tanta,*
" *Pulveris exigni jactu compressa quiescent.*"

VIRGIL.

From its fuccefs in this and feveral other inftances,
which have fallen under my own obfervation, I am
ftrongly inclined to believe, that it will foon become
an object worthy of attention, but muft decline giving
a decifive opinion concerning its fuperiority, till I
know the refult of a variety of cafes, in which it is now
under trial, and alfo receive additional confirmation
from fuch correfpondents, as are lefs prepoffeffed in
its favour than myfelf, becaufe they may be fuppofed
to prove in reality more impartial judges. Therefore
in the interim, I fhall wait with patience for your intended publication, hoping, that by thus collecting the
fentiments and obfervations of feveral practitioners in
different parts of the kingdom, we may at length be
enabled to afcertain the true comparative merit of the
Red Bark, which cannot but afford great fatisfaction
to the public at large, as well as to.

Sir,

Your moft obedient fervant.

A. FOTHERGILL.

Harpur Street, July 6, 1782.

'*A Letter*

A Letter from Mr. EDWARD RIGBY, *an eminent Surgeon, at Norwich.*

DEAR SIR,

I AM very much obliged to you for the honour you have done me, in fending me your Treatife on the Red Peruvian Bark, and I feel myfelf particularly flattered by the manner in which you have folicited an anfwer to your letter.

I am happy that it is in my power to bear the moft ample teftimony to the great and certain efficacy of this Bark ; I have had the fulleft opportunity of trying it in every fpecies of the Intermittent, and have given it to patients under the greateft variety of circumftances refpecting age, conftitution, date of the difeafe, &c. and out of more than a hundred and fifty perfons, who have taken it under my direction, it has not failed in a fingle cafe, ftopping the return of the paroxyfm in the very firft inftance of its application.

I was fortunate enough to receive a fmall parcel of this Bark fo early as October 1781, owing to the friendly communication of Mr. Talbot, Surgeon, at Wymondham, in this county, to whom Mr. Hopkins had fent a few ounces as a fpecimen for trial, half of which he immediately fent to me ; Intermittents being at that time very frequent in Norwich, and its neighbourhood, I had an opportunity of trying it the day after I received it, and I chofe the two worft cafes, which then occurred to me, the one was a Quartan, of many months ftanding, the patient, a boy about eleven years of age, very much worn down by a Quotidian, of nearly as long a date ; the other patient, a young woman of about eighteen years of age, whofe conftitution was alfo much fhattered by the long continuance of the complaint :—to both of them I gave half a grain of Tartar Emetic, at the approach of the next

fit,

fit, which brought off a good deal of Bile from the ftomach, and when the Fever terminated, they began to
take the Bark, an ounce of which was divided into
twelve dofes, all which were taken by each of them
before the times of the expected returns of the complaint ; they both loft the fits, and though they took
no more of it, for they had taken all which was fent
me, they had no return of them until many weeks after, when the weather proved very wet, and then they
went away without taking any more of this, or the
common Bark, for they were both at this time removed to fuch a diftance in the country as not to be able
to fend to me.

My fuccefs in thefe cafes, induced me to procure a
confiderable quantity of it, and as foon as I received
it, I made ufe of it in feveral Tertians then under my
care, in all which it immediately anfwered the intended
purpofe. About this time I had a patient, a gentleman about fifty years of age, who had had a Quartan
more than three months, he had been endeavouring
to cure himfelf by a variety of popular receipts, a-
mongft which was one which contained a confiderable
quantity of the common Bark ; when I firft faw him
it was the day after he had had a fit ; but he was then
much indifpofed, his pulfe was too quick, he had no
appetite, and was much reduced, he took a faline Fever medicine, and an opening draught before the next
fit, at the approach of which, I gave him a grain of
Emetic Tartar, which vomited him confiderably, and
he repeated half a grain of it every three hours, until
the Fever terminated, which, though it lafted a long
while, went off with a greater and more general per-
fpiration than was ufual with him.

I now thought it right to begin giving him the
Bark, but his intermiffion not being quite fo perfect
as I wifhed it, and moreover, my not having yet had
a fufficient number of cafes, in which I had given the
Red Bark, fully to eftablifh its reputation with me, I

thought it moſt prudent to give the common Bark, of which he took two ſcruples at ſuch intervals during the intermiſſion, that he got down more than an ounce and a half of it before the time when the fit was ex-pected, but it did not ſucceed, and he had another fit fully as ſevere as his former one ; I then reſolved to make the next trial with the Red Bark, an ounce of which was taken in the courſe of the ſucceeding inter-miſſion, and in doſes of two ſcruples, and this ſtopped the fit; I prevailed upon him to continue its uſe in the ſame doſe three times a day for a little time, which he did till he had taken two ounces more, but he has taken none ſince; and though his employment expoſes him very much to the weather, and he was ſoon after frequently wet, yet he had not the ſlighteſt return.

From this time I confidently gave it in every caſe which came under my care, and its uſe was invariably attended with the ſame immediate ſucceſs. In the number of thoſe which were cured, were ſeveral whoſe legs were much ſwelled, and their bodies hard, and who appeared to be very rapidly haſtening into a Drop-ſy ; more than twenty of them were children, two were infants, not a year old ; and one, whoſe caſe I ſhall relate, was a ſtriking inſtance of the truth of your remark in the Treatiſe, that the Bark given in conſiderable quantity as near as may be to the time of the approach of the fit, is particularly efficacious.

My patient was a gentleman about twenty-five years of age, robuſt, and of a full habit ; he had had an ir-regular Intermittent about a fortnight, it began as a Quartan, of which he had three fits, it then became a Quotidian, and he had three or four more fits ; I ſaw him about the time when it began to come every day, and I found the Fever was very conſiderable ; the u-ſual treatment, which I need not particularize, was made uſe of, until he appeared to be in a ſtate to take Bark ; the fit after which he was to begin to take it, terminated about ten o'clock at night, and its return

was

was expected the next day between twelve and one at noon. The time of the intermiffion being fhort, he began to take it in dofes of a dram : I had a meffage from him in the night, that it difagreed with him, and that he could not keep it down ; I defired him to per-fevere, but to confider what came up as if he had not taken it, and to fupply its place with another dofe : between five and fix o'clock in the morning he fent to me again, and defired me to go and fee him. I found him much fatigued for want of fleep, which he had been entirely hindered from getting, by his repeated endeavours to take the Bark, every dofe of which, ex-cept the firft, had been rejected, and he feemed con-vinced that his ftomach would not retain it. I pre-vailed upon him, however to take half a dofe, and this kept down ; I ftaid with him fome time, and half an hour after taking the firft half dram, I repeated the fame quantity, which likewife kept down ; I then left him, defiring him to continue it every half hour in the half dofes. Between eight and nine o'clock I was fent for again, and he gave me the fame account of its coming up again as before ; notwithftanding this, I ventured to give him another dofe, which he im-mediately threw out of his mouth, before he had fwallowed any of it ; obferving this, I was more par-ticular in enquiring in what manner he had vomited up, as it was called, the former dofes ; and from the attendant's account I was convinced, that what he ima-gined to have been vomited up, had never been fwal-lowed ; by this unlucky mifmanagement of himfelf, he had taken but two drams and a half inftead of five drams, and there remained five drams and a half to complete the ounce, which was to be taken in little more than three hours ; however, I was determined he fhould perfevere, and I immediately gave him a dram, as I was no longer under any apprehenfion of its coming up, being fully fatisfied that all had been retained which had been actually taken into the
ftomach;

ftomach ; this kept down, and by ftrictly attending
him with a dofe every half hour, or at longeft every
three quarters of an hour, the whole was gotten down
before the time of the expected return of the fit, which
happily prevented its coming, and he has continued
well ever fince. The night having been entirely without fleep, and the patient having been exceffively
fatigued and fretted by his many fruitlefs endeavours
to take his medicine, were very unfavourable circumftances in his fituation, and prevented his Fever from
going off fo perfectly as it did in the former intermiffion, for he was much hotter, and his pulfe quicker
than was to be wifhed, when taking Bark ; however,
the event fully juftified my perfeverance, and ftrongly
proved the efficacy of the medicine.

Many of my patients having been poor and igno-
rant people, and fome of them living at fome diftance
from Norwich, to whom I could only give general di-
rections, as it was impoffible for me to attend them,
you may eafily imagine that moft of that clafs of them
took it carelefsly ; fome of them I know did not take
the quantity prefcribed ; and I recollected one perfon,
a ftrong country girl, about feventeen years of age,
who took an ounce of it at two dofes ; yet all of them
were immediately cured. One patient of this clafs
was a child about eight years old, who had a Quotidian more than two months, and was directed to take
half an ounce of the Cortex between the fit, which was
to terminate on a Monday noon, and was expected to
return on the Tuefday, about the fame time ; on the
following Saturday the child's father called to inform
me that he was cured. Upon enquiry how the child
had taken it, I was furprized to find that he had not
then finifhed the half ounce, for when I told him that
I meant the child fhould have gotten the whole quantity down in one day, he faid he had underftood before, that I directed it to be taken between the Monday and the Saturday night, fo that it was evident the
child

child could not, the firſt day, have taken more than one dram of the medicine, which it is clear, proved ſufficient to ſtop the fit.

From the foregoing, and from ſome other caſes which I have had, as alſo from thoſe communicated to you by my friend Mr. Sherwin of Enfield, one may certainly very ſtrongly preſume, that a much leſs quantity of this Bark than what I have uſually given would anſwer the purpoſe ; it would be a very eaſy matter to aſcertain this by more trials, but at preſent I am not willing to give it in a ſmaller quantity, nor would I recommend it to other practitioners to do ſo, until the reputation of the Bark has been fully and univerſally eſtabliſhed, for whilſt the prejudices of ſome practitioners, who are averſe to new medicines, and the intereſt of ſome Druggiſts, who will be probably ſufferers, by having large ſtocks of the common Bark by them, may in the leaſt degree tend to oppoſe its general uſe, it is to be wiſhed that the teſtimonies in its favour ſhould not only be ſtrong and clear, but that its ſucceſs ſhould be as uniform and invariable as the nature of the medicine admits of, and therefore I would not yet venture to preſcribe a quantity ſo ſmall, as to run a poſſible riſque of its failure, when the quantity I have hitherto uſed, which is an ounce to an adult, and a proportionable leſs quantity to children, has not, with me, in a ſingle inſtance been fallible.

In relating the few caſes above, I did not think it neceſſary to mention the names of the patients, but as the circumſtances which tend to recommend a new medicine cannot have too great notoriety, I will, as they occur to my memory, give you a liſt of ſome of the perſons who have been cured of Intermittents by this Bark, and whoſe ſituations in this neighbourhood render them well known.

Mr. Thomas Smith—Mr. Garland—Mr. Carter, Jun. twice—Mr. Money—Maſter Money, and Miſs Money, of Trowſe—two children of Mr. Barham of

Kirby-

Kirby—Mr. Oliver—Mr. Taylor—Mr. Kiddell's
daughter, of Colney—Mr. Howlett, of Earlham—Servant of John Gay, Efq;—Servant of Mr. Bloom, of
Trowfe—Mr. and Mrs. Clift—Mifs Clift—Son of
Rev, Mr. Anfdell—Mr. Dixon—Mafter Webb--Mifs
Kett--Mafter Bunn--Mrs. Denny of Shottifham--Mr.
Wright of Brecondale--Mrs. Glover of Kirby- Mr.
Smith of Burlingham--Servant of Sir Lambert Blackwell, Bart.

With regard to the common Peruvian Bark, notwithftanding the complaints of its inefficacy have been
great and general, my experience for a few years paft
has given me no reafon to be fo much diffatisfied with
it as I find many others are. I will acknowledge, indeed, that for fome time paft, I have found it neceffary to give a much larger quantity of it than ufual,
and that even when I have given from an ounce and
a half to two ounces of it in fubftance, it has not always fucceeded in ftopping the fit in the firft inftance of
its application, but when I have been able to prevail
upon my patients to perfevere in its ufe, in the fame
quantity, I have never known it fail to ftop the fit
after the fecond intermiffion, in which it has been
taken. I fhould imagine there is no reafon to believe
that the common Bark, which has been ufed for fome
years paft, is not the fame with what was formerly in
ufe, or which was perhaps, originally introduced ; as
far as can be judged by its tafte, and its appearance,
either in the lump, in powder, in decoction, or in any
of the other preparations of it, it feems to me, at leaft,
to be precifely the fame as I have always feen it ; I
have, therefore, never once fufpected that, as a natural production, it has degenerated, much lefs have I
apprehended that any artful means have been ufed by
Druggifts, to render it more faleable, or to increafe
their profit upon it, by which its medical quality has
been diminifhed : Intermittents having been more general in this country for two years paft, than, perhaps,
was

was ever remembered by any Practitioners now living, probably the fame caufe which has made them fo frequent, has made them of a worfe kind, and confequently more difficult to remove ; and to this caufe, rather than to any change in the quality of the Bark, is, in my opinion, to be attributed the late general want of fuccefs in the treatment of this difeafe. Upon the whole then, Sir, from the experience I have had in ufing the two kinds of Bark, which has not been inconfiderable, it is evident to me, that they poffefs the fame medicinal quality, but that the Red Bark has it in a degree greatly fuperior to the pale, which ftrongly favours your fuppofition, that they are both the produce of the fame tree, the Pale or Quill being the Bark of the fmaller branches, and the Red, that of the larger branches, or the trunk of the tree. Having myfelf found fuch fingular fatisfaction in the ufe of this Bark, I fincerely wifh it may be univerfally introduced, and I am perfuaded that every Practitioner, who will give it a fair trial, will immediately prefer it to the Pale. Nothing can be more agreeable in the adminiftration of a medicine, than to be able to fpeak of and foretel its effects with confidence, as it muft be very encouraging to the patient ; this circumftance, and its anfwering the purpofe in a much fmaller quantity than the Pale, are very peculiar advantages which the Red Bark poffeffes ; for in the ufe of the Pale, though, as I before obferved, I make no doubt, but by perfeverance, and taking it in large dofes, it will for the moft part fucceed, yet I have more than once been awkardly fituated with patients, who have taken a large quantity of it without its having anfwered the intention in the firft inftance, I mean in ftopping the fit, after the firft intermiffion, in which it has been given, it not being always an eafy matter to perfuade perfons, under fuch a difappointment, to perfevere taking down a large quantity of a naufeous medicine,

more

more efpecially, when there ftill remain popular prejudices againft the Bark ; and it is a notion received
by fome, that when it does not immediately fuccced it
muft be hurtful.

I have juft received a letter from Mr. Talbot, the
gentleman whom I mentioned before, as having firft
fent me a fpecimen of the Red Bark, with an extract
from which, relative to the fubject, I fhall conclude
this already too long letter ; he informs me, that from
the time he firft made trial of it, he has ufed no o-
ther in Intermittents, that he has given it to more
than fifty perfons, and he has not failed removing the
difeafe in a fingle inflance, though before that time
he had been very unfuccefsful in the ufe of the Pale
Bark. He mentions a cafe, in which four ounces of
the common Bark had been given without effect, and
that an ounce and a half of the Red immediately put
a ftop to the fit :—And he further fays, that he lately
made enquiries about it amongft fome Practitioners in
his neighbourhood, to whom he recommended it, and
that Mr. Swallow of Watton, Mr. Bringloe, of Hing-
ham, Mr. Gibbs, of Buckenham, and one or two
more, have been equally fuccefsful in its ufe as himfelf.

I am, Sir,
With the utmoft refpect,
Your obliged and humble fervant,
EDWARD RIGBY.

Norwich, Sept. 8, 1782.

A Letter

A Letter from Dr. James Maddocks, *Physician to the London Hospital.*

DEAR SIR,

AGREEABLY to your requeſt, I here ſend you an anſwer to the queſtions you propoſed to me, relating to the large and Red Peruvian Bark, lately introduced into uſe in England.

Your firſt queſtion related to my opinion of the medicinal efficacy of this Bark, with that of the paler, ſmaller, and quilled Bark, which for a long ſeries of years has been conſidered in this country as ſuperior to every other ſpecies.

In anſwering this queſtion, the ſhort notice you have given me, and the little time I have at preſent to ſpare, will not admit either of my taking notice of many different kinds of diſeaſes, in which I have had opportunities of obſerving its ſuperior efficacy, or of deſcribing particular caſes of the few diſorders I am to mention; on which I ſhall content myſelf with giving you the general reſult of my obſervations.

The caſes which have afforded me the moſt frequent opportunities of obſerving, and of drawing the moſt ſatisfactory concluſions relative to the ſuperior efficacy of the Red Bark, are thoſe of the Intermittent Fever.

To the beſt of my recollection, it was about the month of October, 1781, when we firſt began to make uſe of the Red Bark, at the London Hoſpital. Immediately after its introduction, the difference, in point of efficacy, between this and the common Bark became very remarkable; inſomuch, that my learned Colleague, Dr. Dickſon, and myſelf, recommended to the Committee of the Hoſpital, at one of their weekly meetings, to purchaſe of the Druggiſt, who had furniſhed the firſt ſpecimen, the whole of his ſtock of

the

the medicine, however great its quantity might be : upon which meafure the Committee, without any he-fitation, immediately refolved.

The London Hofpital is, perhaps, never without a very confiderable number of patients under Intermit-tent Fevers; to which its comparative vicinity to the county of Effex not a little contributes.—When, be-fore the introduction of the Red Bark, we were in the ufe of employing the common Bark upon all occafi-ons, we had found it, in Intermittents, to fall exceed-ingly fhort of that high character for efficacy, which is affigned to it by Dr. Sydenham, and his cotempo-raries.

As to myfelf, I can truly affert, that in the cafes of patients under Intermittent Fevers in the Hofpital, ve-ry feldom indeed was the return of the fit prevented, or even the violence of it much diminifhed at the firft attempt to ftop it, by any quantity of the medicine given in the interval. On the contrary, portions of the Bark for a confiderable length of time, and in very large quantities, were generally neceffary to ftop the progrefs of the diforder, or even to abate its violence ; and on many occafions, from a total want of fuccefs, I have judged it proper to defift from its farther ufe, and to have recourfe to other means of cure.

On the other hand, by the ufe of the Red Bark, I have frequently feen the return of the fit entirely pre-vented upon the firft trial of the medicine given in the interval ; where this is not the cafe, the fubfequent fit is generally lefs violent, and in almoft every cafe the diforder generally difappears in a fhort time.

Several of the cafes in which I have obferved the abovementioned good effects of the Red Bark, were cafes in which the common Bark had previoufly been employed, and continued for a longer or fhorter time, without fuccefs.

With refpect to the out-patients, or fuch as do not refide in, but occafionally come to the Hofpital for

advice

advice and medicines, thefe are much more numerous
than the in patients; among which there occur a great
variety of Intermittents, in all its different types. The
fuperior efficacy of the Red, compared with the common Bark, I have found to be as remarkable in thefe
cafes as in thofe of the in-patients.

Whilft I was in the practice of employing the common Bark, very large dofes of it were generally neceffary to the cure, and the patients ufed to return to the
Hofpital again and again, for repetitions of their medicine; but fince I have ufed the Red Bark, many of
thefe patients have not returned a fecond time. Of
fuch as have returned, fome have informed me that
the dofe prefcribed to be taken during the firft interval had entirely prevented the return of the fit; others,
that the fubfequent fits had been very moderate. And
where the cafes have been otherwife, and I have been
informed of the event, the diforder has given way in a
fhort time.

I fhall trouble you with only one other tribe of difeafes, in which I have had occafion to obferve the fuperior efficacy of the Red Bark. Thefe are periodic
pains, of which difeafes, the periodic Head Ach is the
moft common, and moft generally known.

But I meet pretty frequently with cafes of a fimilar
diforder affecting various other parts of the body than
the head. In fome of thefe, the feat of the pain feems,
to the feelings of the patient, to be fome part of the
parietes of the abdominal, or thoracic cavity, but more
frequently of the former; in others, it feems to be
fome or other of the contents of one of thefe cavities,
but more frequently of the abdominal.

With refpect to the cafes in which the diforder feems
to be feated in fome of the contents of the abdominal
cavity, I have met with many of them, and with fome
that, during the paroxyfm, have been accompanied
with a fharp pyrexia, a moft acute pain, and moft, or
all of the effential, or characteriftic fymptoms of an inflammation

flammation of the viscus ; which, from the place of the pain, seemed to be the seat of the disorder.

These cases, however, differ from inflammations in this, that their paroxysms are succeeded by intermissions, and return at intervals, generally of the Quotidian, sometimes of the Tertian, at other times of less types ; and blood taken from the patient during the utmost violence of the fit, is without the smallest appearance of the size.

In many cases of these disorders, where our view is to prevent the return of the fits, by remedies employed in their intervals, tonics, undoubtedly, are not always proper remedies ; some of them, as, in particular the periodic Head Ach, when it occurs in young and plethoric subjects, may sometimes require the use of evacuants ; but in those cases in which I have judged tonic remedies to be indicated, and have employed the Red Bark, I have found its effects, compared with those of the common Bark, to be similar to those I have above described, respecting Intermittent Fevers.

In support of my opinion of the superior efficacy of the Red Bark in the diseases I have above specified, I have been led from the nature of the subject, to draw my arguments principally from cases of hospital patients, as these, on account of their superior number, afford the better opportunities of comparison ; but the observations I have made in private practice upon these, not to mention other diseases, correspond with and tend to support the conclusion.

You desire to have my opinion relating to the nature of the Red Bark, from what tree it is taken, whether from that which affords the small Quilled Bark commonly in use, or from a different one. In my opinion there can be very little doubt, but that both are taken from the same tree, and that their difference in appearance depends on this circumstance only, that the small or Quilled Bark, is taken either from very young trees, of which all the parts are yet small, or if

ever

ever taken from large, is the produce of their twigs or very fmall branches ; and that on the other hand the Red Bark is taken from well-grown trees, and from their trunks or larger branches.

Certain arguments which I find in your publication on the Red Bark, fome of which are fuggefted by yourfelf, and others, by fome of your correfpondents, are, I think, fufficient to warrant our refting in this conclufion. The principal reafons from which I have been led to adopt it are the following :--

Firft, Becaufe the Red Bark agrees in its fenfible and other qualities, with the fmall Quilled Bark, poffeffing however thefe qualities in a much higher degree.

Second, Becaufe it is very well known, that the peculiar fenfible qualities and powers of moft vegetables, are comparatively little obfervable in the young plants, or tender fhoots.

Third, Becaufe the Tanners know very well from experience, that the Oak Bark which is taken from the trunk or larger branches of the tree, poffeffes much ftronger powers than that taken from the fmaller branches ; and therefore always prefer this in the bufinefs of tanning.

Laftly, What appeared to me upon examining the fpecimens, you lately fhewed me, of Oak Bark, which afforded me an opportunity of comparing the Bark of the trunk, or larger branches of the Oak, with that taken from the fmaller branches, where the Bark of the larger kind appeared of a red hue, and expanded, that of the fmaller, pale and quilled ; a difference exactly fimilar to that we obferve between the two different fpecies of Peruvian Bark.

The only other obfervation I fhall make relating to the Red Bark, is, that when we reflect on the very extraordinary virtues afcribed to the Peruvian Bark, by Dr. Sydenham and Dr. Morton, and particularly on the degree of certainty with which it is by them af-

K   firmed

firmed to have cured Intermittent Fevers ; of which virtues, the fmall and quilled Bark is allowed by all Practitioners, to fall fo very far fhort ; and confider further, that the defcriptions given by the *Materia Medica* writers, cotemporary with the eminent perfons now mentioned, of the Peruvian Bark then in ufe, does not apply to the fmall and Quilled Bark, but does exactly fo to the Red Bark ; and laftly, that the inhabitants of New Spain, and, if I am rightly informed by a gentleman lately arrived from that country, of Old Spain alfo, actually hold the Red Bark in higher eftimation, it muft appear highly reafonable to conclude, that the Red Peruvian Bark is truly the fpecies of this medicine, the virtues of which are fo much extolled by Dr. Sydenham and Dr. Morton, and which was in common ufe with them and their cotemporaries.

To conclude, Sir, I confider the work in which you are engaged as highly commendable. I confider it as one not only of great public utility, inafmuch as it tends to fix the reputation, and extend the ufe of a moft efficacious and important medicine; but alfo as a neceffary one to counteract the endeavours of prejudiced or interefted perfons to oppofe its deferved reputation, and difcourage its general ufe.

I am, Dear Sir,

Your fincere friend,

And humble Servant,

JAMES MADDOCKS.

London, Capel Court, Sept. 20, 1782.

Dr.

Dr. Keir, Phyſician to St. Thomas's Hoſpital, informs me, that in that Hoſpital about 150 lb. of the
Red Bark have been uſed, and he thinks with more
ſuccefs than is uſually experienced from the common
Peruvian Bark.

The employment of it in his practice has not been
confined to Intermittents ; he has alſo uſed it in Mor-
tifications, in Phagedænick Ulcers, in the Convale-
ſcence of Fevers, and in every other complaint that occurred, where the common Bark would have been
deemed a proper remedy.

In oppoſition to the objection ſtated, and refuted by
me, Dr. Keir obſerves, that during the whole of this
extenſive and miſcellaneous uſe of the Red Bark, no
caſe occurred in which there was reaſon to believe any
bad effects to have been produced by it.

Extract of a Letter from Mr. SHIREFF, *an eminent Surgeon and Apothecary at Deptford, in Kent.*

After obſerving that the ſituation of Deptford, and
its environs, renders the inhabitants of that village extremely ſubject to Intermittent Fevers of a very obſti-
nate nature, ſome of which he found more difficult to
cure, than even ſuch as he had ſeen on the weſt coaſt
of Sumatra, where they put on a more formidable appearance than in Europe. He proceeds as follows :

" From the ſeveral patients whom I attended, I ſe-
lected the three following caſes to try the effects of the
Red Bark ; not to enumerate every particular, I ſhall
only obſerve, that in all of them the common Bark
had failed in a ſingular manner. Each of them had
ſuffered frequent relapſes, the firſt caſe eſpecially ; ſhe
had ſcarce any reſpite for nine months in Lincolnſhire ;
ſhe had removed to this place to try the effects of a
different air ; before I ſaw her ſhe had been here for
three months, without finding any benefit.

C A S E

C A S E I.

' A gentlewoman of a weak conftitution, and naturally of a nervous habit of body, had been feized laft autumn with a fimple Tertian in Lincolnfhire; upon her removing to this place it had affumed the Quartan type, and was of three months ftanding; having found very little relief from Bark and other remedies, fhe had declined calling in any affiftance; but her hufband, alarmed at her extreme weaknefs during a paroxyfm, fent for me : I was informed of the above particulars, and found her with an icteric countenance, fwelled ancles, and other marks of great debility : after fome difficulty, I prevailed upon her to take medicines; I fent her immediately feveral dofes of Red Bark, each containing only one fcruple, on account of her naufeating every thing that was prefented her, defiring her to begin after a general moifture had come on, with an abatement of thirft and head ach, and to be repeated every four or fix hours; fhe continued the medicine in this manner, for four or five days, and as the fubfequent paroxyfm had been more mild than the one preceding it; I could not prevail upon her to take the Bark fo frequently; fhe however continued its ufe for fourteen days longer, each day taking four fcruples, which entirely removed her complaints—fhe is now in perfect health.

C A S E II.

A young gentleman, naturally of a robuft and healthy habit of body, had fuffered feveral paroxyfms of a double Tertian to attack him, without ufing any method to prevent them; it was his determination to truft to nature for a cure, rather than take fuch large, and frequent dofes of the Bark, as he had fome months before done, without any permanent effects.

But

But a delirium feizing him in one of the paroxyfms, his relations fent for me at midnight: I found him fenfible, his body covered with a moft profufe fweat, and loaded with bed-cloaths, the curtains drawn clofe, and the external air carefully excluded from the room ; having removed every obftacle to the free admiffion of air, and his body wiped with a dry cloth, and in place of hot drinks, made ftill hotter with fpice, I or-dered toaft and water, acidulated with lemon, and o-ther diluting liquors, to be given him almoft cold ; I procured his confent to make one more trial of medi-cine; accordingly, half a dram of Red Bark was giv-en him immediately, and repeated every four hours ; he miffed the next period, and after continuing his medicine three days longer, only three times in the day; he left it entirely off;—he has fuffered no re-lapfe, and is now in health.

C A S E III.

A gentleman, after expofing himfelf to a damp e-vening, was feized with the common fymptoms of Fe-ver ; an emetic was given immediately, and followed by a laxative, not apprehending that he was attacked again with an Intermittent, he continued to go to Lon-don ; at the expected period, however, he was again taken ill, and the paroxyfm was rendered very fevere, by his imprudently walking home after it had com-menced. When the febrile fymptoms abated, and the fkin became moift, I gave him half a dram of the Red Bark, with orders to repeat it every three hours ; he fuffered no return, and now remains well ; being much expofed to the weather, I have advifed him to conti-nue fmall dofes of it twice in the day.

I am, Dear Sir,
Your obliged and humble Servant,
J. L. SHIRREFF.

Deptford, Sept. 14, 1782.

A Second

DEAR SIR,

A S medical attention has of late been defervedly engaged on the fubject of the Red Peruvian Bark, and as the public are not a little interefted in the refult, I take liberty to fubmit to your confideration, a few more curfory obfervations, which have occured fince my laft.

The teftimony which you have already produced from fo many refpectable Practitioners, who could be under no temptation, either to conceal its failures, or to exaggerate its virtues, renders it unneceffary to add any frefh evidence of its fuperiority ; otherwife I might mention fome late inftances of its fuccefs, in certain inveterate Agues, which had entirely baffled the ordinary Bark. Therefore, whatever doubts or difficulties may be now raifed concerning its identity with the Bark formerly ufed by Morton and Sydenham, can by no means invalidate the facts which have been advanced in fupport of its real efficacy. That it was, however, in actual ufe about the beginning of the prefent century feems demonftrable.

In the year 1702, the cargo of Bark which was captured on board a Spanifh galleon, a parcel of which fell into the poffeffion of Mr. Pearfon, an eminent apothecary in the city only four years ago, appears, from every circumftance, to have been no other than the drug now under confideration. But what feems ftill more worthy our attention, is, that after the fpace of about 78 years, it fhould ftill afford a much ftronger decoction than that of the common Bark, and alfo furpafs it in the cure of fevers, and other difeafes : an evident proof that this Bark retains it medicinal powers much longer than could have been imagined. In further

further confirmation of this fingular property, and alfo of its early ufe in this country, allow me to tran-
fcribe a remarkable paffage from Dr. Lifter, who
mentions fome of its moft charadteriftic marks : " Pro-
" pria experientia teftor, me ante 20 annos cortice
" trunci fæpe ufum effe ad craffitiem, & latitudinem
" volæ manus, magnis & profundis fulcis, & fiffuris
" confpicuo, velut in vetufto arbore, imo eundem ali-
" quando cariofum ; & olim, & nunc, vix unquam
" fruftravit eventu optimo, & defiderato, maxime fi
" ejus modus, et tempus exhibitionis rite obfervan-
" tur *." Add to this, a ftill further proof with
which I have been lately favoured by Dr. Smith, a
very ancient phyfician near Andover, who affures me,
that having obtained a fample of the Red Bark, he immediately recognized it, " both by the fmell, tafte and
" colour, to be the fame that was commonly ufed
" fifty years ago." To which he fubjoins fome recent inftances of its fuccefs in Intermittents, which
had refifted the ordinary Bark.

From the year 1640, that the Peruvian Bark was
firft imported into Spain, its reputation increafed till
the old unpeeled trees becoming fcarce, the inhabitants of Loxa, mixed other Barks with it, which being detected, it fell into fuch difcredit, that, in the
year 1690, feveral chefts of it lay in the warehoufes at
Piura, and nobody to purchafe it. From this circumftance, and from the infignificant dofes in which it was
adminiftered, it difappointed the public expectation fo
much, as to be generally difcarded, till Tabor, an ad-
venturous Englifh practitioner, by giving more ade-
quate dofes of the genuine drug, revived its reputation ; when its fame fpread fo rapidly, that the Spanifh
merchants, at length, found it difficult to fupply the
demand of their cuftomers for full grown Bark, and
therefore partly through neceffity, and partly through
political

* De Hydr. p. 56.

political œconomy, fubftituted the fmall Bark with which they have long furnifhed the European markets. Hence may be explained, why they now affect to extol the Quill Bark, which is more eafily prepared, and more readily obtained, in almoft any quantity, and that without deftroying the trees. M. Condamine, who vifited Loxa, about fifty years ago, affures us, that the Red Bark was allowed to furpafs the other forts, but was grown, even then, extremely fcarce, on account of the reafon already affigned *.

Of late years Peruvian Bark has become fuch an important article of commerce, that our merchants are glad to procure fuch as is offered ; but no candid Spanifh Practitioner, who has tried the different forts, will, it is prefumed, be at a lofs in determining to which the real preference ought to be given.

As the prefent ftock of genuine Red Bark cannot but be extremely difproportionate to the demand, it only remains, that we earneftly admonifh younger practitioners not to be too precipitate in drawing unfavourable conclufions from the refult of their prefent trials, but to fufpend their judgment, till a frefh fupply fhall enable them to pronounce with more certainty, concerning its comparative powers.

As it hitherto promifes to be much fuperior to the common Bark, in the fpeedy cure of Intermittents, it will alfo behove them to ufe the greater circumfpection in afcertaining the true nature, and tendency of the difeafe, viz. whether it is a primary, or only a fecondary affection, whether certain obftacles are not previoufly to be removed ; whether the cortex is not contraindicated ; and laftly, whether the fudden fuppreffion of periodical motions may not prove productive of fome more dangerous derangement in the fyftem.

From

* Mem. de l'Acad. des Sc. 1738.

From the prefent indifcriminate ufe of the Peruvian
Bark, in difeafes fo diametrically oppofite in their nature, I cannot help thinking, that the inactivity of this
univerfal Catholicon, fo generally lamented of late,
has been rather a fortunate circumftance, and that the
inertnefs of the remedy has often prevented a feries of
evils, which muft have enfued from fuch a prepofter-
ous abufe of the genuine drug.

In Intermittents which are purely idiopathic, and
proceed from an epidemic conftitution of the atmof-
phere, without any concomitant difeafe, or internal inflammation, the Bark may generally, without hefita-
tion, be freely exhibited ; and in highly urgent cafes
of this kind, which prevail in marfhy countries, and
fultry climates, wherein the remiffions are very fhort,
this medicine can fcarcely be adminiftered too foon,
or too liberally. Under fuch hazardous circumftances
time is too precious to admit of preparatory evacuations, and I moreover concur with you in confidering
them as frequently unneceffary, if not injurious.

On the contrary, it muft be allowed, that Agues are,
fometimes merely fymptomatic of fome other more
dangerous affection, and ought to be confidered by the
attentive Practitioner, as remedies, rather than difeafes.
Thus in the gout, the pain, inflammation, and tume-
faction of the toe, is not the principal difeafe, but a
critical *metaftafis*, in order to its cure ; fo febrile paroxyfms are, in certain cafes, to be confidered as the
falutary efforts of nature, to fubdue fome morbific
caufe, or to remove fome confirmed difeafe, of a more
fatal tendency. In fuch cafes, nature is to be affifted,
not difarmed of thefe ufeful weapons, by which fhe
fometimes combats Palfies, Epilepfies, and other Her-
culean maladies, which all the artillery of medicine
could otherwife never have fubdued.

This being accomplifhed, the febrile paroxyfms either fubfide fpontaneoufly, or may be now fafely removed by this powerful febrifuge.

On

On the other hand, there are not wanting inftances, where the fupervening Ague is fo far from removing the former difeafe, that it ferves but to exafperate its fymptoms, and if fuffered to continue, to produce ftill other dangerous affections. To determine with accuracy and precifion in thefe different fituations, demands a degree of medical difcernment and deliberation, which but too feldom occur in the hurry of modern practice.

With refpect to the general operation of the Peruvian Bark, I entirely coincide with what you have fo fatisfactorily advanced, and fhall only prefume to fubjoin the following reflections.

Phyficians, in attacking Putrid Fevers and other obftinate difeafes with the Bark, feem extremely folicitous to impregnate the whole mafs of fluids with its fpecific virtue, yet excellent as it is, when applied to the nervous furface of the alimentary canal, nature feems to me, never to have intended that it fhould enter the blood, and has therefore wifely placed firm barriers to prevent its admiffion into the interior parts of the machine. A fubftance which is capable of undergoing repeated macerations, and decoctions in water for many months, without being wholly divefted of its bitternefs and aftringency, could not eafily be fubdued in the blood-veffels; but would probably prove (at leaft in its native ftate) utterly incompatable with the laws of the fyftem. Dr. Friend accordingly informs us, that no fooner had he injected two ounces of a decoction of this medicine into the jugular vein of a dog, than it produced fevere palpitations, convulfions, and death *. The Provident Guardian of the human frame, thus kindly checks the wild career of afpiring mortals, when, through the mifts of boafted fcience, they blindly purfue devious paths that often lead to

dangerous

---

* Emmealog. c. xiv.

dangerous errors. Not that our refearches into the laws of the œconomy, and the operation of medicines can be too deep, or profecuted with too much ardour, fo long as we follow the clue of accurate obfervation, and draw no conclufions but what are fairly deducible from the phenomena; but unfortunately, from the little we know of thefe matters, we often prefume a great deal concerning the major part which remains unknown, and the mifconceptions which thence enfue in theory, are transferred into practice.

In the late German war, the French army on their return from Bohemia, were feized with Tertian Agues of the putrid kind, which at length terminated in critical abceffes, which formed behind the ears, and in the arm-pits. When thefe abceffes were fully matured, they were opened according to the ufual mode of practice in fimilar cafes. But no fooner was this unfortunate operation performed, than the fymptoms recurred, accompanied with extreme proftration of ftrength, under which the patients generally funk in a few days. But when the fick were left to Nature's own management, without any attempt to promote, or retard fuppuration, or to open the abceffes, the purulent matter was fpontaneoufly difcharged by the inteftinal canal, or fome of the other emunctories. The practitioners were now led to acquiefce in the mode of cure pointed out by Nature, and from this time, almoft all who were affected with the difeafe recovered. *

Thus Art often boldly ufurps the province of Nature, and undertakes to regulate the inordinate motions of a complicated machine, and not unfrequently by very improper, or very inadequate means. From this fource proceed innumerable errors in the treatment of difeafes, and endlefs miftakes concerning the effects of medicines. Intermittent Fevers, and the Bark, the fubjects now under difcuffion, afford pregnant examples of both. Nor can any reformation be

expected

* Mem. de l'Acad. des Scien. de Stockholm.

expected till more attention is paid to that excellent maxim of the illustrious Verulam :

"*Non fingendum, aut excogitandum, fed inveniendum quid* NATURA *faciat, aut ferat.*"

I remain, Dear Sir,

Your moft obedient Servant,

A. FOTHERGILL.

London, Sept. 20, 1782.

APPENDIX

APPENDIX

A Letter from Dr. SAMUEL FOART SIMMONS, F. R. S.
to Dr. SAUNDERS.

DEAR SIR,

THE fuperior efficacy of the Red Bark is now
fo clearly eftablifhed, that it would be fuperflu-
ous to trouble you with a detail of the numerous cafes
in which I have tried it. In the cure of Intermittents,
fome of them very obftinate ones, and that had refifted
the Common Bark, it has not once failed me; and I
have lately feen an infufion of it remove a double Tertian of three months ftanding, in a young woman, who
had taken a large quantity of Oak Bark, and of the
pale Peruvian Bark, both in decoction and in fubftance,
without experiencing any relief from either. A lady
upwards of fixty years old, who refides in a part of
Kent where Agues are very frequent, and who for more
than a twelvemonth had laboured under a Quartan,
which had brought on fymptoms of Jaundice, and refifted the Common Bark, change of fituation, and a
variety of other remedies, likewife owes her recovery
to the Red Bark. My learned and worthy friend, the
celebrated

celebrated Profeffor Camper, informs me, that he has experienced the fame good effects from this Bark in Friefland, a country where agues are endemial. From repeated experiments, he is convinced that fix grains of it are equal in efficacy to a fcruple of the pale Peruvian Bark. Now that the virtues of this excellent remedy are fo fully afcertained, you will naturally be defirous of enquiring more particularly into its hiftory. From the largenefs of this Bark, you were at firft inclined to confider it as the Bark of the trunk, or larger branches of the *Cinchona Officinalis*, *Lin.* and the Quilled Bark as a production of the twigs, or fmaller branches of the fame tree ; but having lately met with fome very good Red Bark, as fmall as the Quilled Bark in common ufe, you are now, it feems, difpofed to think, that the tree which produces it may be a variety, or perhaps a diftinct fpecies of the *Cinchona Officinalis.*— On this head I have fome intelligence to communicate, which I am perfuaded will be acceptable to you.

Amongft the papers of the late M. Jofeph de Juffieu, (brother of the famous Bernard de Juffieu) one of the French Academicians, who went to Quito in Spanifh America, in order to afcertain the figure of the earth, and who died lately at Paris, feveral interefting obfervations have been found relative to the Peruvian Bark. Thefe have been communicated to the Royal Medical Society at Paris, by his nephew Dr. Anthony de Juffieu. In his defcription of the genus, M. de Juffieu agrees with his fellow traveller, M. la Condamine, but he admits a greater number of fpecies. Thefe, however, may perhaps be very properly reduced to two, as the reft feem to be only varieties.

The firft fpecies includes the red, the yellow, and the knotty (*le noueux*) Barks, all of which have very fmooth leaves, flowers of a purplifh colour, and inodorous, with a Bark that is bitter to the tafte, and more or lefs coloured. Of thefe three the Red is held in the higheft eftimation, and it is this fort of Bark, according

ing

ing to M. de Juffieu, which was employed in the early
days of this remedy in Europe, and which acquired it
fo much, and fuch deferved celebrity. The tree that
produces it is become fo exceeding fcarce, that in the
year 1739, M. de Juffieu found it growing only in a
few places in the neighbourhood of Loxa, fo that the
inhabitants of Peru had been obliged to fubftitute the
yellow and knotty Barks in its ftead, both of which
they are faid to prefer for their own ufe, becaufe they
fuppofe them to be lefs active and heating. But M.
de Juffieu, who had experienced the good effects of
the Red Bark, both in his own perfon, and in others,
confidered it as infinitely fuperior to the reft. Even
the trees that produce the yellow and knotty Barks are
faid to be diminifhing in number fo faft, that it is to
be feared they will in time become extinct, unlefs a regular mode of cultivating them is adopted, or they
are difcovered elfewhere.

The fecond fpecies includes the White Barks, of
which there are four varieties. They have all of them
broad roundifh hairy leaves; the flowers are red, very
odoriferous, and furnifhed with hairs on their infide
furface. The fruit is longer than that of the former
fpecies, and the outer Bark is of a whitifh colour. In
two of thefe varieties, the inner layers of the Bark are
of a reddifh hue; they have a flightly bitter tafte, and
when frefh, are faid to poffefs a flight febrifuge quality, but which they foon lofe. The Bark of the other
two is entirely white, infipid, and of no efficacy.

M. Ant. de Juffieu has ftill in his poffeffion fome
extract prepared by his uncle upwards of forty years
ago at Loxa, from the Red Bark. Some trials lately
made with it, prove it to be infinitely fuperior in efficacy to the extract of Bark in common ufe, fo that its
virtues do not feem to have been diminifhed by keeping.

M. de Juffieu, in his travels, found a few of the trees
that produce the yellow and knotty Barks, growing in
different

different parts of the valley that extends along the
chain of the Andes, and in the diftrict of Yungas, which
is near it; but it was only about Loxa, in the 4th deg.
of S. Lat. that he faw forefts of thofe trees. It would
feem therefore, that the heat peculiar to fuch a latitude
is more genial to the *Cinchona* than that of any other
climate, and of courfe we can hope to meet with it only in fuch a temperature. Upon this principle we
might be tempted to look for it at a fimilar diftance
from the equator in a northern latitude. This has actually been done : Don Cafimir Ortoga, Profeffor of
Botany at Madrid, has lately by order of the Spanifh
Minifter for the American department, fent to the
Royal Medical Society at Paris, fpecimens of two fpecies of *Cinchona* recently difcovered in America, in the
province of Santa-Fe, which is fituated 4 deg. and half
of north latitude.

Thefe fpecimens are well preferved, but not quite perfect, as the flowers are wanting. The leaves and fruit
of one of thefe fpecies exactly refemble thofe of the
Red Bark, fent by M. la Condamine, from Peru, and
which are ftill preferved in M. de Juffieu's *Hortus Sic-
cus*. The other fpecimen proves to be a White Bark,
and of courfe a bad fpecies. The Spanifh Minifter
accompanies thefe fpecimens with a requeft, that the
Society would inform him what degree of attention
they merited. The Society have of courfe given his
Excellency every neceffary information on this fubject,
and as he is now aware of the great importance of the
Red Bark, there can be no doubt but proper directi-
ons will be given for its cultivation in *Santa-Fe*, not
only on account of its fcarcity at Loxa, but becaufe
it will be much more eafily conveyed to Europe, as a
river that runs through the province of *Santa-Fe* empties itfelf into the harbour of Carthagena, fo that we
may hope foon to fee a new fource opened for this
admirable remedy.

I cannot

I cannot conclude this long letter, without thank-
ing you for the pleafure and information I have re-
ceived from the perufal of your ingenious publication,
which forms fo valuable an addition to the Materia
Medica. Believe me, with great truth,
Dear Sir,
Your faithful and obliged Friend
and humble Servant,
SAM. FOART SIMMONS.
Air-Street, Piccadilly, Feb. 17, 1783.

Extract of a Letter from Mr. AIKIN, *Surgeon, at Warrington.*

THE fpecimen of Red Bark that I examined,
agreed perfectly with your defcription. I tried its
ftrength by the tefts of the action of water and fpirit,
and was immediately convinced of its poffeffing a much
larger fhare of active matter than the beft Pale Bark ;
the moft decifive experiment was the quantity of refin
obtained by evaporating a fpirituous tincture drawn
from equal quantities of both forts ; that yielded by
the Red Bark, was in proportion of three to two of
that extracted from the Pale Bark, and yet on infufing
the woody *refidua* of each in boiling water, that of the
Red gave a bitter liquor, which ftruck a manifeft
black with martial vitriol, whereas, that of the Pale,
gave out neither fapid nor colouring particles. My
friend, Dr. Haygarth, informs me, that on a fimilar
experiment made at Chefter Infirmary, the refin extracted from Red Bark was, to that from Pale, as 229
to 130.
With refpect to the medical efficacy of this Bark, I
am not able to relate any fair comparifon of it, with the
common fort, as Agues, in which the beft opportunity
is afforded for fuch a comparifon, have been uncommon

here since I attended to this subject ; I have prescri
ed it in most of the cases in which Peruvian Bark i.
usually given, and from the general result, I have n
doubt of its virtues being similar, but probably superior, to those of the kind commonly preferred.

I am, Dear Sir,

With sincere Regard,

Your obedient Friend and Servant,

J. AIKIN.

Warrington, Feb. 10, 1783.

I HOPE I shall not be considered as presumptu-
ous, in concluding, that the foregoing observations are
sufficient for determining the superior efficacy of the
Red Peruvian Bark. And it seems reasonable to hope,
that the introduction of this kind of Bark may be attended with the happiest effects, and enable us to op-
pose more successfully those malignant and remittent
fevers of warm climates, and unfavourable situations
so destructive to our fleets and armies.

I cannot, however, finish, without returning my best
thanks to the gentlemen who have favoured me with
their correspondence ; and I think it necessary to express
my obligations to many gentlemen whose letters were
too late for publication.

They all concur in recommending the Red Peruvian Bark, as more efficacious and powerful than any
other kind.

From the numerous trials I have made with it, in
Intermittent Fevers, and other diseases, I am disposed
to conclude, that it need be employed only in half the
quantity we generally recommend of other Bark.

I have likewise the satisfaction of assuring my readers, that it is now in general use in all the large Hos-
pitals in London ; and such is the preference given to
it, that the demand is difficultly supplied.

Be

Be careful in the choice of it, by attending to the characters which diftinguifh it from the large Bark, hitherto rejected by our Druggifts.

I fhall continue to be diligent in my enquiries on the fubject, and I moft earneftly requeft the favour of my friends, that they will perfevere in fupplying me with accurate obfervations, fo as to determine, with precifion, in what other difeafes this valuable remedy may be ufed with fafety and advantage.

The operations and effects of remedies can only be afcertained by the united induftry and experience of intelligent men ; who, by being aware of the difficulty of making obfervations, are fufficiently guarded againft the fallacy to which they are unavoidably expofed.

THE END.

THE GENUINE

RED BARK,

EXAMINED AND APPROVED BY

DOCTOR SAUNDERS,

May be had (neat as imported) of

MR. OLIVER SMITH,

AND

MR. WILLIAM SCOLLAY,

DRUGGISTS, in CORNHILL, BOSTON;

Where alfo may be had, the beft ASSORTMENT of

DRUGS AND MEDICINES.

THE FOLLOWING

NEW AMERICAN

PUBLICATIONS,

Are to be Sold at the loweſt Prices by WILLI-
AM GREEN, Bookſeller, at Shakeſpeare's
Head, oppoſite Bromfield's Lane, Marlborough-
Street, Boston:

THOUGHTS in PRISON, in Five Parts,
viz. The Impriſonment—The Retroſpect—Public
Puniſhment—The Trial—Futurity. By the Rev.
William Dodd, L. L. D. To which are added, his
laſt Prayer, written in the night before his death,
and other Miſcellaneous Pieces.

——Theſe evils I deſerve, and more :
Acknowledge them from God inflicted on me
Juſtly ; yet deſpair not of his final pardon,
Whoſe ear is ever open, and his eye
Gracious to re-admit the Suppliant ! MILTON.

⁎ If aught in Nature can touch the niceſt ſprings of
the benevolent mind with tender compaſſion, and
draw forth the tears of real ſenſibility, it muſt be,
When a good Man falls. The unfortunate Author
of

of this Work, in this laſt Performance, at the near and certain approach of an ignominious death, diſcovers, as it were, the very inmoſt receſſes of his ſoul without diſguiſe.

LETTERS on the IMPROVEMENT of the FEMALE MIND. By Mrs Chapone.

This excellent performance, which has already gone through four editions in Europe; and in which the Author has happily found the art of conveying the moſt important inſtructions in the moſt elegant language, calls for the attention of every young lady who wiſhes either to be improved or entertained, and when the influence of the fair ſex on the manners of the age is duly conſidered, it will readily be granted, that Mrs. Chapone's Letters, form one of the moſt uſeful as well as pleaſing modern publications.

FRIENDSHIP in DEATH: In Twenty Letters from the Dead to the Living. To which are added, Moral and Entertaining Letters, in Proſe and Verſe. In Three Parts. By Mrs. Elizabeth Rowe.

The character of the author is too well eſtabliſhed to require the aid of the panegyriſt. The beauty and elegance of the language can only be exceeded by the excellence of the ſentiments. She has plucked the choiceſt flowers from the garden of virtue and placed them on the head of merit, while ſhe has dreſſed vice in ſable attire. This book is well calculated for the inſtruction and amuſement of the Ladies.

ADDRESSES

ADDRESSES TO YOUNG MEN.
By JAMES FORDYCE, D. D.

Thofe who have read Dr. Fordyce's Sermons to young women, will naturally form great expectations from his Addreffes to young men, and we may venture to affure them that their expectations will not be difappointed. The tender and affectionate concern which the author expreffes for the improvement of youth in knowledge and virtue, and in every thing that can render them truly amiable, fhews the goodnefs of his heart, and the agreeable and entertaining manner in which he communicates to them the moft ufeful inftructions, does honour to his tafte and genius.

AN ESSAY on the Caufes, Nature and Cure of CONSUMPTIONS. In a Letter to a Friend. By JOHN MOORE, M. D.

The juftly deferved reputation of Dr. Moore's letters, will entitle this, on a diforder unhappily very prevalent in our country, to particular attention, as it is wrote in a ftyle not only elegant, but free from all thofe technical terms which often render the writings of Phyficians unintelligible to any but fuch as make the medical art their peculiar ftudy, and as it has long been agreed by the moft eminent in the profeffion, that notwithftanding the generally received opinion, of the conftant fatal termination of Confumptions, they may, by early and clofe attention to proper regimen, be fometimes if not often cured.

THE ART OF SPEAKING: An Effay; —In which are given. I. Rules for expreffing properly the principal paffions and humours, which occur in reading or public fpeaking :—And, II. Leffons exhibiting a variety of matter for practice.
The

The advantages of Elocution, whether in impreffing the important truths of religion from the facred Defk, in the Senate, or at the Bar, muft be apparent to every perfon of the fmalleft tafte for letters. The difpofition evident in our rifing republic, of cultivating every talent, which may either adorn human nature or benefit mankind, have induced the publifher of this edition of the Art of Speaking, to offer to his countrymen a performance, calculated not only for the improvement of youth, of whatever calling or profeffion in life defigned, but from its mafterly execution, promifing pleafure and improvement to the moft enlightened mind.

THE COMPLETE WORKS of LAURENCE STERNE, A. M. Prebendary of York, and Vicar of Sutton on the Foreft, and of Stillington, near York; with the Life of the Author, his Letters to Eliza, &c. &c. in Five Volumes.

An Affortment of B O O K S, are conftantly kept for fale at faid Shop; where Book-binding is alfo performed at a reafonable rate, with neatnefs and difpatch.

Account Books, Memorandum Books, Children's Books, Writing Paper, Ink, Ink Powder and Cake Ink, Black lead Pencils, Sealing Wax, Wafers, Leather Ink Holders and Ink Stands: Together with moft Articles in the Stationary Way, may be had at the above Place, at the loweft Rate for Cafh.